Environmental and Economic Evaluations of Building Energy Retrofits

Environmental and Economic Evaluations of Building Energy Retrofits

Editors

Cynthia Hou
Joseph H.K. Lai

MDPI • Basel • Beijing • Wuhan • Barcelona • Belgrade • Manchester • Tokyo • Cluj • Tianjin

Editors
Cynthia Hou
The Hong Kong Polytechnic University
China

Joseph H.K. Lai
The Hong Kong Polytechnic University
China

Editorial Office
MDPI
St. Alban-Anlage 66
4052 Basel, Switzerland

This is a reprint of articles from the Special Issue published online in the open access journal *Energies* (ISSN 1996-1073) (available at: https://www.mdpi.com/journal/energies/special_issues/Building_Retrofits).

For citation purposes, cite each article independently as indicated on the article page online and as indicated below:

LastName, A.A.; LastName, B.B.; LastName, C.C. Article Title. *Journal Name* **Year**, *Volume Number*, Page Range.

ISBN 978-3-0365-4551-6 (Hbk)
ISBN 978-3-0365-4552-3 (PDF)

Contents

About the Editors

Cynthia Hou

Cynthia Hou has been engaged in research on facilities management in both the private and public sectors. Her areas of research focus include post-occupancy evaluation, workplace management, strategic facilities management, and heritage adaptive reuse and revitalisation. Recently, she has leveraged smart technologies in her research by investigating smart hotels, smart heritage facilities management, and virtual reality (VR) technologies applications in built environment education. She seeks to investigate efficient and sustainable built environment management strategies from a facilities management perspective. She is also keen on adopting digitalisation approaches in her research, with the aim of delineating the dynamics among people, facilities, and space in the built environment and exploring data-driven solutions for facilities management problems in society. Cynthia is a member of the Royal Institution of Chartered Surveyors and the programme leader for the Master Programme in Facilities Management, The Hong Kong Polytechnic University

Joseph H.K. Lai

With a first-class honours degree in building services engineering, a master's degree in environmental management, a PhD in facility management (FM) and over a decade of professional practice, Dr Joseph Lai combines professional and academic roles as a scholar in FM teaching and research. Dr Lai is Associate Head of the Department of Building Environment and Energy Engineering, Faculty of Construction and Environment, The Hong Kong Polytechnic University. The honours Dr Lai has received include the Teaching Excellence Award, Faculty Teaching Excellence Award, Dean's Award for Outstanding Achievement in Academic Programme Development, and Distinguished Educator Award of the International Facility Management Association. Dr Lai is Co-Editor of the Facilities journal and Associate Editor of the *Smart and Sustainable Built Environment* journal.

energies

MDPI

Article

Transaction Cost and Agency Perspectives on Eco-Certification of Existing Buildings: A Study of Hong Kong

Yung Yau [1], Huiying (Cynthia) Hou [2], Ka Chi Yip [3] and Queena Kun Qian [4,*]

1 Department of Sociology and Social Policy, Institute of Policy Studies, Lingnan University, Hong Kong 999077, China; yungyau@ln.edu.hk
2 Department of Building Environment and Energy Engineering, The Hong Kong Polytechnic University, Hong Kong 999077, China; Cynthia.hou@polyu.edu.hk
3 Institute of Policy Studies, Lingnan University, Hong Kong 999077, China; kachiyip2@ln.edu.hk
4 Faculty of Architecture and the Built Environment, Delft University of Technology, 2628 BL Delft, The Netherlands
* Correspondence: K.Qian@tudelft.nl

Abstract: Eco-certification schemes are usually launched with various incentives provided by local governments to facilitate green building development and building energy retrofits. A number of barriers to building energy retrofitting have been identified in previous literature, while the barriers to the eco-certification of existing buildings are under-researched. Drawing on a set of building data retrievable from the BEAM Society and other sources, we carried out an analysis and found the building energy retrofitting, as well as the certification process, were unwelcomed in multi-owned residential buildings. The identified shortfall is put forward from the perspectives of transaction cost theory and agency theory. The findings reveal that high transaction costs incurred during negotiations and coordination among a large number of co-owners within a typical apartment building can outweigh the benefits of retrofitting and eco-certification. Besides, the remuneration structure of third-party property management agents discourages agents from facilitating co-owners to initiate retrofitting. This study provides significant implications for policymakers to understand the concerns of building owners and managers over the decisions and the processes of both the building energy retrofits and eco-certification. The problems and barriers unveiled in this study will facilitate the refining of current energy efficiency policies and related incentives designs.

Keywords: building energy performance; building energy retrofits; green building certification; transaction costs; agency theory; incentives

Citation: Yau, Y.; Hou, H.; Yip, K.C.; Qian, Q.K. Transaction Cost and Agency Perspectives on Eco-Certification of Existing Buildings: A Study of Hong Kong. *Energies* **2021**, *14*, 6375. https://doi.org/10.3390/en14196375

Academic Editor: Antonio Gagliano

Received: 1 September 2021
Accepted: 27 September 2021
Published: 5 October 2021

1. Introduction

In 2019, building construction and operations accounted for 35% of final energy consumption and 38% of energy (and process-related) emissions [1]. The latest global energy consumption data suggest that building energy use remains a significant proportion of overall energy demand. The percentages of building energy consumption against total final energy consumption in different areas are shown as follows: 57% in Africa, 40% in Europe, 26% in ASEAN, China, and India, and 24% in Central and South Africa [1,2]. Many governments across the world have made significant efforts in promoting energy efficiency in their built environments, providing various initiatives to drive building owners or operators to retrofit their existing buildings. The United States has launched and planned 203 energy-efficiency related policies for the building sector, followed by Australia and Canada, with 94 energy efficiency policies, respectively [3]. China has launched and planned 236 energy efficiency policies, with 59 being relevant to the building sector, varying from regulatory instruments, building codes and standards, minimum energy performance standards, economic instruments, strategic planning, etc., to fiscal/financial incentives.

To promote energy efficiency initiatives in Hong Kong, where buildings account for 90% of electricity consumption, the Hong Kong government has adopted both statutory and non-statutory approaches to enhance the city's overall building energy performance. In Hong Kong, Leadership in Energy and Environmental Design (LEED) and BEAM Plus (a green building rating tool managed by the Hong Kong Green Building Council) are two major environmental building rating and certification schemes. They have been well received by the construction industry in Hong Kong. In the past decade, the two voluntary schemes have significantly facilitated green building development and fostered building energy reduction in Hong Kong. The two schemes include a set of rating systems for the design, construction, operation, and maintenance of buildings and neighbourhoods, based on which, the building owners have their buildings/projects certified to obtain eco-certification. Among the 42,000 existing buildings in Hong Kong, 1.5% (approximately 630 buildings) have accredited or registered to the BEAM plus certification system, and 0.34% (approximately 142.8 buildings) have accredited or registered to the LEED certification system [4]. In recent years, two main concerns for the Hong Kong government—to achieve energy reduction targets of 40% in the built environment by 2025—involved enabling existing buildings toward energy retrofitting, and obtaining certification under energy-efficiency related schemes [5]. In 2019, BEAM Plus New Buildings (BEAM Plus NB) and BEAM Plus Existing Buildings (BEAM Plus EB) were launched to foster a sustainable built environment in Hong Kong.

It is easier to have an aged building eco-certified if it has undergone energy retrofitting. However, in reality, the reasoning and decision-making processes for existing building energy retrofitting and eco-certification can be complicated. For example, the barriers to eco-certification of existing buildings would affect the decisions of building energy retrofits, especially when the building owners' decisions are motivated by the incentives brought by the eco-certification scheme [6]. In this light, investigating the patterns of eco-certification of existing buildings would help deepen the understanding of the barriers toward building energy retrofits. In this study, an inductive approach is adopted to examine the characteristics of the existing buildings that are eco-certified. The factors that hinder implementing the building energy retrofits and the eco-certification are identified and discussed from a transaction cost theory and an agency theory perspective. The findings of this study will help policymakers understand the concerns and behaviours of building owners regarding the process of eco-certification application, enable building owners and operators to evaluate the costs occurred by the barriers, and make rational decisions to overcome possible barriers in the future.

The paper is organised as follows: Section 2 provides a literature review on the barriers to building energy retrofits and the eco-certification of existing buildings. Section 3 introduces the methodology of this study followed by a detailed presentation of the findings in Section 4. Section 5 elaborates the discussion based on the findings and Section 6 presents the conclusion of this study.

2. Context of the Research

In practice, local governments support eco-certification by launching various incentives in both statutory and non-statutory forms (e.g., policy incentives, tax incentives, funding assistance, etc.) [6]. Furthermore, eco-certification organisations also integrate incentives into the design of the eco-certification schemes and their certification processes, such as reducing the cost for volume certification, creating different application pathways, etc. The benefits of these incentives are attractive to developers and building owners and, to a certain extent, serve as a huge motivation for them to apply for eco-certification for new projects.

The barriers to green building development have been profoundly discussed in numerous studies. Darko and Chan [7] provided a comprehensive review of the barriers to green building adoption and identified 37 barriers in the literature. Arguably, these barriers also hinder the eco-certification of new buildings. Perspectives derived from the transaction cost theory (TC) [8,9] and cost–benefit analysis [10–12] have been used to analyse the developers' behaviour and the development process of green building development. Moreover, eco-certification of new dwellings may not be considered because homebuyers may pay little attention to the eco-certification or energy labels in their home purchase decisions [13,14].

Yet, the extant literature does not distinguish the barriers to green building development and barriers, to obtain eco-certification for buildings. In other words, discussion on the barriers to eco-certification of new buildings is usually mixed, regarding the barriers to green building development. Even worse, there has been a dearth of literature on the barriers to eco-certification and energy retrofits of existing buildings. The factors that incentivise or dis-incentivise existing buildings' eco-certification and energy retrofits could be different from those associated with new buildings. Therefore, it is worthwhile to study the factors that shape decision making for eco-certification and energy retrofits in our existing building stock.

The focus of this study is to investigate the possible barriers to building energy retrofits through examining the issues that affect the eco-certification of existing buildings based on the database of eco-certification of existing buildings in Hong Kong. Thus, the target of the investigation is the management of existing buildings, such as the decision on whether to implement building energy retrofits and apply for eco-certification, the selection of the certification pathway, etc.

2.1. Barriers to Building Energy Retrofits and Eco-Certification of Existing Building

Building energy retrofitting works generally involve "replacements, modifications, and refurbishments of existing buildings to enhance the energy efficiency, conservation, and savings" [15]; the minimization of energy consumption and the maximization of economic benefits are the two prime objectives of retrofitting. Building control improvement and building component implementation are two major energy-retrofitting strategies to increase energy efficiency and reduce the energy demand of the building [16]. The process of building energy retrofits involves multiple stakeholders from different professional backgrounds and with different intentions towards the retrofitting decisions. Among the involved stakeholders, building owners play an important role in building energy retrofitting and eco-certification decision-makings. A number of barriers regarding the stakeholders' perceptions towards building energy retrofits, as well as their expected outcomes, are identified in the literature. For example, building owners may have an aversion to energy efficiency refurbishment measures because of the lack of interest in energy efficiency issues, financial means, long-term perspectives, and trust towards contractors [17]. As far as municipalities are concerned, factors, such as unawareness about the energy problem, difficulties with goal setting and data collection, and lack of expertise in the municipalities to analyse the data and develop an effective plan, hinder government-led energy retrofit projects [18].

Hong et al. [19] studied the commercial building energy retrofitting projects in China and suggested that a lack of expertise and resources to identify and evaluate cost-effective energy retrofit strategies are major barriers for owners when it comes to pursuing energy retrofitting. Hou et al. [20] identified a number of issues that would decrease building owners' willingness to retrofit their buildings, including unclear stakeholder obligations, difficulties in coordinating multiple parties, and complexity of retrofit implementation. Aside from individual research projects by academic scholars, international associations also put effort into identifying the barriers to building energy retrofitting. Building Performance Institute Europe identified four barriers to building energy retrofits, categorizing them into four categories: (1) financial; (2) institutional and administrative; (3) awareness, advice, and skills; and (4) separation of expenditure and benefit [21]. Climate Policy Initiative also identified four barriers to building energy retrofits, namely: (1) embryonic markets; (2) lack of information; (3) misaligned financial incentives; and (4) undervaluing energy efficiency [22].

As a significant number of existing buildings are aging and facing an urgent need to upgrade their operational, economic, and environmental performance worldwide, the pursuit of green certification for existing buildings has become an inevitable trend [23]. Aktas and Ozorhon [24] only focused on the green building certification process and identified three major barriers to existing the eco-certification process of buildings in developing countries—unavailability of approved materials, poor design of buildings, and difficulties with the documentation process. One of the findings from their study was that the building owners do not perceive the cost as a barrier to 'green' their existing buildings. The possible reason is that the building owners (owners of commercial buildings) see the certification as an opportunity to enhance their corporate image. Thus, they are more flexible with the budget on green implementation of existing buildings.

2.2. *Building Energy Retrofits and Eco-Certification in Hong Kong*

2.2.1. Statutory Regulations and Policy Incentives for Building Energy Retrofits

In Hong Kong, the Building Energy Efficiency Ordinance (BEEO) (Cap. 610) was formulated to enforce certain prescribed types of buildings to comply with building energy codes (BECs) and/or the energy audit code (EAC). In addition, the Hong Kong Energy Efficiency Registration Scheme for Buildings (HKEERSB) was introduced to recognise buildings that outperform the statutory requirements under the BEEO. The HKEERSB was officially launched in 1998 in order to promote the adoption of the BECs by providing the certification to a building complying with one or more of the BECs [25].

In 2018, the Hong Kong government launched a tax incentive scheme to further encourage building owners to pursue the application of HKEERSB: as long as the building owners have their buildings certified by BEAM Plus (managed by the Hong Kong Green Building Council) or other internationally recognised building environmental assessment systems, such as LEED, they are eligible to apply for relevant tax deductions [26,27]. Building owners and building managers are familiar with the BEEO. The newly launched tax incentives have drawn the attention of policymakers to the barriers of building energy retrofits and the eco-certification process for existing buildings in Hong Kong.

2.2.2. Background of the BEAM Plus

Building Environmental Assessment Method (BEAM) Plus is the prevailing rating tool for green buildings in Hong Kong. BEAM Plus, conceived in 1996 as a voluntary private sector initiative, has developed into an internationally recognised green building rating tools for new buildings (NB), existing buildings (EB), interiors (BI), for shops, office, retails, and neighbourhood (ND) [28]. To achieve the target set out in the Energy Saving Plan by 2025, the Hong Kong Green Building Council (HKGBC) issued the new version of BEAM Plus Existing Building (BEAM Plus EB) V2.0 in 2016 [5,29]. It includes the new assessment framework with two certification pathways: a comprehensive scheme and a selective scheme. In order to have the building certified under the comprehensive scheme of the BEAM Plus EB, the building performance shall be assessed under all seven aspects, including management, energy use, indoor environment, water use, materials and waste aspects, site aspects, and innovation and additions. The applicant can also choose to have the building performance assessed under one or more specific aspects through the selective scheme pathway. Based on the assessment scores by BEAM professionals, the buildings are awarded certain ratings (e.g., platinum, gold, silver, and bronze) that reflect the actual performance of the building. The four-level rating system applies to both a comprehensive scheme and a selective scheme [29,30].

In order to promote and facilitate the certification of BEAM Plus EB, a volume certification mechanism was introduced to provide a faster and more economical manner for certification application. The applicants can choose to have all buildings or multiple buildings certified by a portfolio assessment mechanism in one go, at a lower cost, through a volume certification approach. This study focuses on the buildings that are certified under the BEAM Plus EB and analyses their background information, including the application mechanism (comprehensive or selective scheme), the certification rating, the building ownership status, the building/estate development information, and the geographical information.

2.3. A Transaction Cost (TC) and Agency Perspectives towards Building Energy Retrofits

Transaction cost (TC) theory has been widely used to analyse the externalities that occur within a firm, in the interaction between a firm and the market, and the process of public policies implementation [31–33]. Transaction costs impact economic performance because high transaction costs can lead to failure of an institutional arrangement [34,35]. Prohibitively high transaction costs tend to inhibit collective actions [36,37]. "Transaction cost" has been discussed and measured in numerous studies on public policies to promote green buildings [7,8,10,38]. It is the main idea of these works that certain types of TCs would occur with different stakeholders in the process of development, certification, and management of the green construction project, and these TCs would undermine the policy's effectiveness and implementation efficiency. It is commonly found that some common barriers exist during the implementation of green building development-related policies, such as information gathering [39,40], internal and external negotiation [39,41,42], innovative technology acceptance, and adoption [43,44]. These studies conceptualised the identified "barriers" with TCs and used TC theory to transform the vague phenomenon into a tangible concept. Qian et al. [8] adopted the transaction cost theory with support of the empirical data from expert interviews to examine the cost and benefits (both actual and hidden ones) in the process of implementing the gross floor area (GFA) concession incentive scheme in Hong Kong. With a similar approach, Fan et al. [9] used the transaction cost theory to measure three dimensions of transaction costs involved in the scheme implementation, namely asset specificity, uncertainty, and frequency. With the theoretical base on transaction cost theory, Fan et al. [11] conducted a case study using empirical data to support the hidden costs and benefits of the same scheme.

Agency theory is often used to explain and resolve disputes over priorities between principals and their agents. The difference in agreement between the principals and agents during the transaction process leads to agency problems. The agency problems generally result from the conflicting goals of the principals and agents, intensified by the information asymmetry [45,46]. The principal–agent dilemma may also stem from moral hazards and adverse selections. A moral hazard occurs when an agent attempts to make a profit on a contract because the principal is unable to observe the agent's behaviour after entering into the contract [47]. On the other hand, uncertainty concerning an agent's characteristics and preferences prior to creating a contract could lead to an adverse selection [48].

In the past decade, agency theory has been increasingly employed to analyse green management strategies in organisations [49–51]. Yet, studies using agency theory in investigating energy efficiency are still rare. Kumbaroğlu and Madlener [52] attempted to analyse building energy efficiency problems, focusing on investigating the benefits and conflicts of interests between investors and users based on agency theory. Liang et al. [6] adopted agency theory to explain the agency problems in energy-efficiency retrofits and developed a principal–agent model to map out the problems of two sets of principal–agent relationships: government vs. building owners and building owners vs. tenants. They argued that "incentives" play an important role in the benefit–cost analysis in energy retrofitting decisions. The model illustrates the economic relationships among the three stakeholders (government, building owners, and tenants) under four scenarios. In building energy retrofitting and the eco-certification process, property management companies play the 'agent's' role, implementing the tasks (building energy retrofitting and eco-certification) for the building owners, who play the 'delegator' role, according to the principal–agent relationship. Although both sides in the contractual relationship receive certain incentives from the eco-certification schemes, their interests appear to be imbalanced under certain circumstances. This study aims to identify the possible causes that hinder property management companies from facilitating building owners to implement building energy retrofits and gain eco-certification.

3. Research Methodology and Data

An indicative approach was adopted for this empirical study. Secondary data about eco-certification under BEAM Plus were first collected from various sources. The data were then consolidated and analysed to see if there were any specific patterns regarding eco-certifications of the existing building in Hong Kong. Possible explanations were then tendered to explain the patterns. In the stage of data collection, a list of certified existing buildings was captured on 21 May 2021, from the HKGBC BEAM Pro Project Directory. (https://www.hkgbc.org.hk/eng/beam-plus/beam-plus-dir-stat/BEAMPlusDirectory.jsp, accessed on 21 May 2021) Only the valid and certified BEAM Pro existing buildings projects available on the online project directory on the captioned date were selected for investigation. In total, the project directory contained 193 existing building projects and 201 valid BEAM Pro existing building certifications. The number of building projects and certifications did not reconcile because a few building projects obtained more than one BEAM Plus certification in BEAM Plus V2.0 (Selective Scheme).

Figure 1 outlines the data collated for desk study with their sources. We principally examined three main categories of information, including (i) Hong Kong BEAM eco-certification figures (certification scheme, project rating, certification year); (ii) background information of each BEAM Plus certified building; and (iii) property governance matters (e.g., ownership status and owner's information). After the data collection, all relevant data were compiled into a project-specific database for further processing. The skeleton of this project-specific database was primarily connected to two local green building databases—HKGBC's BEAM Project Directory and BEAM Society Limited's BEAM Plus Certified Building Database. We also amassed other necessary project data, such as the completion year and the property management company, in diverse sources. Trustworthy databases managed by the government sectors, property consultant agencies, and the local property management professional bodies were utilised. In addition, we reviewed the websites and publications, such as annual reports and environmental, social, and governance (ESG) reports of the property owners and management companies to ensure information accuracy. All information had undergone further validation by cross-checking to ensure data precision.

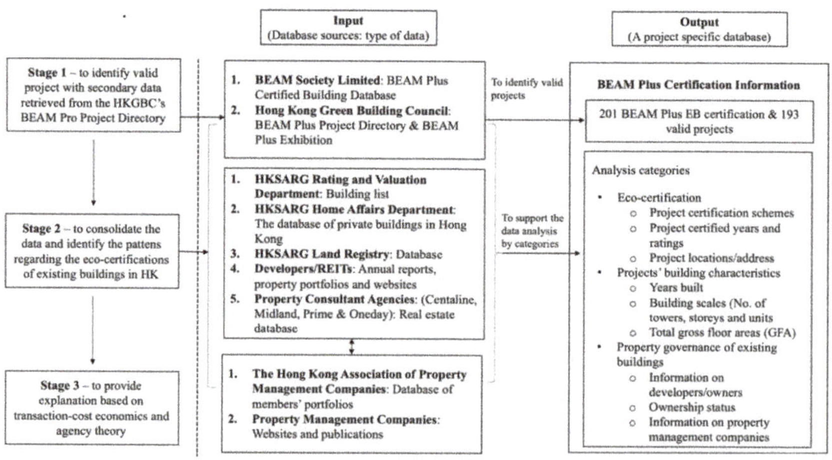

Figure 1. Research design, data retrieving, and analysis process.

To prevent ambiguity, the term "buildings" in the results section refers to "building projects" (the building or a group of buildings) with respect to BEAM Plus certification records.

4. Results

This section will present the analytical results of the eco-certification existing buildings in Hong Kong.

4.1. By Certification Schemes and Ratings

Table 1 shows that only 41 buildings in Hong Kong are completely certified with six sustainable building areas. (This figure aggregates 40 building projects that achieved BEAM Plus comprehensive certification and an NGO headquarter, which gradually achieved all six performance aspects through BEAM Existing Buildings Version 2.0 Selective Scheme.) Approximately 20% of the certifications were recognised through the comprehensive scheme, while more than 80% of BEAM EB certifications were awarded in the selective scheme. Most BEAM-certified existing buildings were rated in the management aspect only. This kind of certification contributed to nearly 90% of the certifications in selective schemes and 71% of all types of certifications (including comprehensive scheme and selective schemes). Although energy use and site aspects were the second and third most prevalent aspects rated in the selective schemes, they respectively contributed to 6.2% and 2.5% of selective certifications across six sustainability fields. Excluding an outlier project where the headquarters of a non-governmental organization (NGO) was certified in all six aspects of the selective scheme, i.e., equivalent to a comprehensive certification, no selective certification was recognised in the area of material and waste aspects or indoor environmental quality. The proliferation of management certifications also skewed the overall distribution of the green EB certification rating, as shown in Table 2.

Table 1. Breakdowns of certifications by schemes (N = 201).

Certification Scheme	No. and Percentage	Aspect	No. and Percentage
BEAM EB Version 1.1 and Version 1.2	10 (5.0%)	-	-
BEAM EB Version 2.0 (Comprehensive Scheme)	30 (14.9%)	-	-
BEAM EB Version 2.0 (Selective Scheme)	161 (80.1%) [1]	Management	143 (88.8%) [2]
		Site	4 (2.5%)
		Materials and Waste	1 (0.6%)
		Energy Use	10 (6.2%)
		Water Use	2 (0.6%)
		Indoor Environmental Quality	1 (0.6%)

[1] ()—the percentage of participated scheme in respect of all BEAM Plus certifications. [2] ()—the percentage of participated aspect in respect of BEAM EB Version 2.0 Selective Scheme.

Table 2. Breakdowns of certifications by pathways and ratings (N = 201).

Rating	No. via Individual Certification	No. via Volume Certification	Overall
Final platinum/excellent	57 (85.1%)	10 (7.5%)	67 (33.3%)
Final gold/very good	4 (6.0%)	0 (0%)	4 (2.0%)
Final silver/good	3 (4.5%)	124 (92.5%)	127 (63.2%)
Final bronze/satisfactory	3 (4.5%)	0 (0%)	3 (1.5%)
Total	67 (100%)	134 (100%)	201 (100%)

In general, over 97% of BEAM existing buildings were awarded a silver/good rating or higher. For the volume certification of the property portfolio submitted by the two major developers (Sun Hung Kai Properties and Link REITs), and their subsidiary property management companies, more than 60% of the certifications are classified into "Final Silver/Good". A small proportion of applications (1.5%) are assigned to a satisfactory classification.

On the other hand, the final ratings of the projects through the two certification pathways are different. There is a significant relationship between the final rating and certification pathway (χ^2 = 149.54; $p < 0.01$). If solely considering the final rating by individual certification, it is revealed that around 85% of the certifications are given with the highest rating (final platinum or excellent). It implies that applicants applying local green building schemes through individual certification may focus on the sustainability dimension qualities (maximum credits achieved). Yet, why do the individually certified have higher ratings? Do project types and building uses also affect BEAM certification methods? Moreover, as it is hypothesised that buildings with single ownership pay fewer transaction costs for green building retrofitting, what are the implications of different ownership statuses on BEAM EB certifications? We analyse property management governance of BEAM-certified buildings by their ownership status, building uses, and owner types.

4.2. By Building Ownership Status, Project Types, and Sectors

The previous section showed that the majority of BEAM-certified existing buildings were assessed only based on the aspect of sustainable building management. It is equally important to examine the ownership status, building type, and sector in unpacking the potential barriers of sustainable building retrofits in Hong Kong. Generally speaking, developers, property owners, and property management companies are imperative in regard to building governance and sustainable building retrofit. Collective actions are necessary for co-owners to initiate building energy retrofits. From the transaction costs and agency perspectives, it is expected that existing buildings in multiple ownerships are less likely to be eco-certified. This is because high transaction costs are usually incurred in the negotiations and coordination among co-owners of multi-owned properties when it comes to initiating certification or improving common areas of buildings. Besides, co-owners in multi-owned properties often need to rely on property management companies in eco-certification or energy retrofit projects. The companies may not act in the best interests of the co-owners.

Table 3 breaks down the certifications by ownership status, project types (or building uses), and owner types. Regarding ownership status, it is found that nearly 72.5% of certified existing buildings in Hong Kong are owned by a single owner (hereafter: single-owned), while the rest (27.5%) are owned by multi-owners (hereafter: multi-owned). Regarding the certification pathways—a difference in terms of ownership status can be seen. The ownership status ratio (i.e., number of certified single-owned buildings to that of certified multi-owned buildings) surged to about nine to one (90% single-owned versus 10% multi-owned) if cases where volume certification was excluded. (Contrasting ownership status outcomes are observed if we separate the samples concerning certification pathways. Regarding ownership status per application type, around 35% of projects via portfolio certifications are multiple-owned (whereas 63.6% are single-owned correspondingly). By contrast, about 11.7% of projects via individual certifications are multiple-owned (and 88.3% of projects are single-owned).) This implies that more single-owned buildings applied the BEAM certification via individual application than via volume application.

By project type, nearly 70% of BEAM-certified existing buildings are commercial properties, while 15% are residential buildings. Industrial buildings, government/institution or community (GIC) buildings, and buildings of other types only account for approximately 8%, 6%, and 2% of the whole sample, respectively. All 15 industrial buildings were certified through volume certification through the selective scheme ("Management" aspect) by a major developer. The figures in Table 3 indicate that the BEAM-certified existing buildings in Hong Kong are predominantly single-owned commercial use. As far as the owner type is concerned, more than 92.7% of the eco-certified projects belong to the private sector, whereas the public organizations or NGOs own less than 10%. There is a significant relationship between the project type and sector (owner type) (χ^2 = 168.42; p < 0.01). The reason behind such findings is straightforward. The GIC projects predominately belong to the public sector while nearly all commercial, residential, and industrial projects are from the private sector.

Table 3. Cross-table showing the breakdowns of certified projects by project types, ownership status, and owner types (N = 193).

Project Type/Ownership Status		No. of Projects				Percentage of Projects			
		Public	NGO	Private	Overall	Public	NGO	Private	Overall
Commercial	Total	0	0	133.5 [4]	133.5	0%	0%	100%	100%
	SO [1]	0	0	119	119	0%	0%	100%	89.1%
	MO [2]	0	0	14.5	14.5	0%	0%	100%	10.9%
Residential	Total	1	1	27.5	29.5	3.4%	3.4%	93.2%	100%
	SO	1	1	1	3	33.3%	33.3%	33.3%	10.2%
	MO	0	0	26.5	26.5	0%	0%	100%	89.8%
Industrial	Total	0	0	15	15	0%	0%	100%	100%
	SO	0	0	3	3	0%	0%	100%	20.0%
	MO	0	0	12	12	0%	0%	100%	80.0%
Government/Institution or Community (GIC)	Total	10	2	0	12	83.3%	16.7%	0%	100%
	SO	10	2	0	12	83.3%	16.7%	0%	100%
	MO	0	0	0	0	0%	0%	0%	0%
Other Types [3]	Total	0	0	3	3	0%	0%	100%	100%
	SO	0	0	3	3	0%	0%	100%	100%
	MO	0	0	0	0	0%	0%	0%	0%
All	Total	11	3	179	193	5.7%	1.6%	92.7%	100%
	SO	11	3	126	140	7.9%	2.1%	90.0%	72.5%
	MO	0	0	53	53	0%	0%	100%	27.5%

[1] SO: single ownership; [2] MO: multiple ownership; [3] others included freight forwarding centres, data centres, and technology parks. [4] Some projects count as half (0.5) for a particular property use. For example, a mixed-use project (residential-cum-commercial project) counts as 0.5 for "commercial" and 0.5 for "residential".

Moreover, a significant relationship is found between the sector (owner type) and certification pathway ($\chi^2 = 34.28$; $p < 0.01$). Moreover, projects certified via volume certification have an average score significantly higher than those certified individually (t-statistics = 15.08; $p < 0.01$). That means volume certification generally results in less superior eco-labels. Table 4 enumerates the BEAM-certified projects by the project's key owners or property developers. Sun Hung Kai Properties has the greatest number of BEAM-certified existing buildings projects (75 projects in total or 38.8%), followed by Link REIT (50 projects or 25.9%), Swire Properties (13 projects or 6.7%), as well as Nan Fung Group (12 projects or 6.2%). However, 96% of Sun Hung Kai Properties projects were granted through the volume certification pathway under the selective scheme ("management" aspect). Similarly, all projects managed by Link REIT and Nan Fung Group were certified via volume certification under the selective scheme ("management" aspect). For Swire Property, HKSAR Government and Hongkong Land, have the most BEAM-certified existing buildings through the individual certification pathway citywide. For the Swire Property and Hongkong Land, all their certified projects are single-owned Grade A commercial buildings with the highest performance grading (Final Platinum). The records of these buildings were traced against the HK-BEAM system (HK-BEAM certification is the oldest version of BEAM tool). It reveals that all of them were previously certified in either/both HK-BEAM new or/and existing buildings, and nine of them were HK-BEAM-accredited new buildings with the highest rating (Platinum) (certified between 1996 and 2005). In other words, these BEAM-credited existing buildings are either pre-existing green buildings certified by HK-BEAM certification before, or buildings that were managed in a sustainable manner in previous years. Single ownership and engagement of a subsidiary property management agent facilitate the sustainable building retrofitting and eco-certification process.

Table 4. Breakdowns of certified projects by developers or property owners (N = 193).

Property Owner/Developer (Parent Organization) [1]	HKGBC Patronship	No. of BEAM-Certified Existing Buildings			Average Rating Score [2]
		via Individual Certification	via Volume Certification	All	
Public Sector		**11**	**0**	**11**	**1.46**
HKSAR Government	-	8	0	8	1.38
University of Hong Kong	Marble	1	0	1	1.00
Vocational Training Council	-	2	0	2	2.00
NGO		**3**	**0**	**3**	**1.67**
Business Environment Council	-	1	0	1	1.00
Hong Kong Housing Society	-	1	0	1	3.00
Tung Wah Group of Hospitals	-	1	0	1	1.00
Private Sector		**45**	**134**	**179**	**2.43**
CK Asset Holding	-	1	0	1	1.00
Ever Gain Plaza Management	-	1	0	1	2.00
Gammon Construction	Marble	1	0	1	1.00
Great Eagle Holdings	Silver	1	0	1	1.00
Hang Lung Group	Gold	4	0	4	1.25
Henderson	Gold	1	0	1	1.00
HKEX	-	1	0	1	1.00
Hongkong Land	Gold	6	0	6	1.00
Hysan Development	Gold	2	0	2	1.00
Link REIT	Gold	0	50	50	1.67
Mapletree	-	1	0	1	3.00
MTR Corporation	-	1	0	1	1.00
Nan Fung Group	Gold	0	12	12	2.00
New World Development	Platinum	3	0	3	3.00
Pacific Century Premium Developments	-	1	0	1	1.00
Paramatta Estate Management	-	1	0	1	1.00
Shui On Group	-	1	0	1	4.00
Sino	Gold	3	0	3	2.00
Sun Hung Kai Properties	Gold	4 [3]	72 [3]	75	2.64
Swire	Platinum	13	0	13	1.00
Total		60 [3]	134 [3]	193	–
Average Rating Score		2.86	1.25	2.36	–

[1] Regarding Joint Venture Project or ownership status with complicated situations, this table regards the major developers as (1) the company who carried out BEAM certification as main developers if equally shared; (2) the one with the largest ownership share. [2] This refers to the arithmetic mean of BEAM Plus ratings of all projects in the portfolio of a particular organization, with 1 = final platinum/excellent; 2 = final gold/very good; 3 = final silver/good; and 4 = final bronze/satisfactory. [3] The energy use certification of Sun Hung Kai Centre was obtained through the individual application while the management certification was through volume application. We count this project in both certification pathways.

We should note that that many applications for certifications were lodged by the property management agents rather than the property owners. Furthermore, there are vigilant affiliations between the owners/developers of the certified buildings and the property management agents managing the buildings. Among the BEAM-certified existing buildings owned by the private sector, 99% (178 out of 179) of the BEAM-certified buildings are developed and managed by companies that belong to the same groups. In many cases of multi-owned private buildings, the property management companies concerned are subsidiaries of the developers. For single-owned buildings, the building owners often dedicate the property management tasks to their own specialised in-house property teams or subsidiary property management companies. Table 5 enlists the building owners (or developers) and property management companies that have close relationships.

Table 5. Reciprocal relationships between developers and property management companies.

Property Owners/Developers (Parent Company)	Property Management Company (PMC)	No. of Projects under the Same Group
CK Asset Holding Limited	Goodwell Property Management Ltd.	1
Gammon Construction	Gammon Construction Ltd.	1
Great Eagle Holdings	Keysen Property Management Services Ltd.	1
Hang Lung Group	Hang Lung Properties Ltd.	4
Henderson	Henderson Sunlight Property Management Ltd.	1
HKEX	Hong Kong Exchanges and Clearing Ltd.	1
Hongkong Land	Hongkong Land Group Ltd.	6
Hysan Development	Hysan Property Management Ltd.	2
Link REIT	1. Link Asset Management Ltd. 2. Link Property Management Services Ltd.	50
Mapletree	Mapletree North Asia Property Management Ltd.	1
MTR Corporation	MTR Corporation Ltd.	1
Nan Fung Group	1. Hon Hing Enterprises Ltd. 2. Main Shine Development Ltd. 3. Mount Nicholson Property Management Ltd. 4. Nan Fung Property Management New Charm Management Ltd.	12
New World Development	Urban Property Management Ltd.	3
Pacific Century Premium Developments	Island South Property Management Ltd.	1
Paramatta Estate Management	Paramatta Estate Management Ltd.	1
Shui On Group	Shui On Centre Property Management Ltd.	1
Sino Group	Sino Estates Management Ltd.	3
Sun Hung Kai Properties	1. Hong Yip Service Co. Ltd. 2. Kai Shing Management Services Ltd. 3. Royal Elite Service Company Ltd. 4. S.H.K. Real Estate Management Co. Ltd. 5. Sun Hung Kai (Harbour Centre) Ltd. 6. Supreme Management Services Ltd.	75
Swire	Swire Properties Management Ltd.	13
Total		178

4.3. By the Building/Estate Development Scale

The development scales of certified projects of residential buildings can be measured with two indicators: (i) the number of residential units; and (ii) the gross floor areas (GFAs) of the projects assessed by BEAM professionals. Two classification methods with equal intervals and Jenk's natural breaks were adopted to estimate the number of residential units, and the estimation was undertaken using the geospatial thematic classification with the aid of the software QGIS 3.

Table 6 shows the breakdowns of BEAM-certified projects for residential uses by the number of residential uses using two different intervals. For the 29 BEAM-certified existing buildings for residential uses, a total of 25,096 residential units were identified, comprising about 0.85% of the total housing stock in Hong Kong (the total number of residential units in Hong Kong as at 2020 was about 2,924,000 [43].) However, most of the residential buildings were certified under the selective scheme with regard to the management aspect, and their certification applications were driven by the property management companies. Among the 29 samples, one project, a senior staff quarters tower for a government-funded university (renovated in 2016), was assessed under the comprehensive scheme, while 4 (out of 29) residential projects were certified via the individual certification pathway.

By Jenk's natural break classification, over one-third of the BEAM-certified projects had less than 265 residential units. Three projects were low-rise luxury terrace-type residences, and another seven were single/twin high-rise towers. Furthermore, one large-scale estate (City One Shatin) was certified via individual certification under the selective scheme ("Management" aspect). In this project, the management company only selected 3 out of 52 blocks in the estate (536 out of 10,642 units or 5.3%) for BEAM Plus certification application. The possible reason behind this might relate to the collective decision-making issue, which is that the agreement among residents was difficult to seek. This situation may affect the decision-making of green building retrofitting in the estate. This helps explain why no multi-owned large-scale housing estates are awarded or registered for BEAM eco-certification by their individual applications. This phenomenon will be further discussed in Section 5.

Table 6. Number of residential units of BEAM-certified projects for residential uses (N = 29).

Equal Interval		Natural Break (Jenks)	
No. of Residential Units	**No. (%) of BEAM EB Residential Projects**	**No. of Residential Units**	**No. (%) of BEAM EB Residential Projects**
10–916	20 (69.0%)	10–264	11 (37.9%)
917–1823	5 (17.2%)	265–723	9 (31.0%)
1824–2729	2 (6.9%)	724–1159	4 (13.8%)
2730–3636	0 (0%)	1160–2771	3 (10.3%)
3637–4542	2 (6.9%)	2772–4542	2 (6.9%)
Overall	29 (100%)	Overall	29 (100%)

Unlike their residential counterparts, many non-residential projects often have an open-plan design. It is thus impracticable to compare project scales solely based on the unit numbers. Yet, it is still envisaged that an existing building with a larger GFA is likely to have more owners or tenants than the one with a smaller GFA. Therefore, the number of the owners or tenants involved in a project can be roughly estimated by the assessed GFAs. As indicated in Table 7, about one-third of the 155 BEAM-certified projects are small-scale developments. Five projects are classified as developments with an exceptional or mega-large scale.

Table 7. Assessed GFAs of BEAM-certified projects (N = 155) [1].

Project Scale [2]	Corresponding Construction Floor Area (sq. m.) [2]	No. (%) of BEAM Plus EB Projects	No. (%) of Projects Achieving BEAM "Energy Use" Performance [3]
Extra small (ES)	≤2499	2 (1.3%)	2 (1.3%)
Small (S)	2500–24,999	49 (31.6%)	10 (6.4%)
Medium (M)	25,000–49,999	37 (23.9%)	8 (5.2%)
Large (L)	50,000–99,999	40 (25.8%)	15 (9.7%)
Extra large (EL)	100,000–199,999	22 (14.2%)	7 (4.5%)
Mega large (MG)	200,000–400,000	4 (2.6%)	2 (1.3%)
Exceptional Scale	>400,000	1 (0.6%)	0 (0%)
Overall		155 (100%)	44 (28.4%)

[1] The information of assessed GFAs of the certified projects in the database of the BEAM Society Ltd. was last updated in early 2020. By then, the information of assessed GFAs was available for 155 out of 193 certified projects only. [2] The categorization of the project scale follows the scale adopted in the determination of the BEAM Plus application fee. Due to limited available clarification on the BEAM's assessed GFA, we assume that the construction floor area (CFA) equals the assessed GFA. [3] Projects achieving "energy use" performance refer to those projects certified under the comprehensive scheme or selective scheme ("energy use" aspect).

4.4. By Regions and Districts

Table 8 presents the breakdowns of certified projects by building uses and regions. Figure 2 shows the geographical distribution of BEAM-certified projects by building uses (in dots of different colours) and the proportions of projects certified via individual certification pathways in each district. Figure 2 indicates a high concentration of the certified projects in and around Hong Kong's central business district (CBD). This can be explained by the findings in Section 4.2, which is that most of the certified projects are commercial projects. BEAM-certified single-owned Grade A office buildings are clustered in the CBD areas where Central, Admiralty, Wan Chai, and Tsim Sha Tsui districts are located. On the other hand, from Figure 2, we can also find certified projects in different districts. This can be explained by the practices of volume certification adopted by the large developers or landlords whose green properties in their portfolios scatter citywide.

Table 8. Breakdowns of BEAM-certified projects by building uses and regions.

Region	% of Projects via Individual Certification (N = 60)						% of Projects (All Pathways) (N = 193)					
	C	R	I	GIC	OU	All	C	R	I	GIC	OU	All
Hong Kong Island	73.2%	40.0%	0%	8.3%	0%	55.0%	30.5%	40.7%	14.3%	8.3%	0%	29.0%
Kowloon	19.5%	40.0%	0%	58.3%	0%	28.3%	30.5%	30.5%	50.0%	58.3%	0%	33.2%
New Territories	7.3%	20.0%	0%	33.3%	100%	16.7%	39.0%	28.8%	35.7%	33.3%	100%	37.8%
Overall	100%	100%	0%	100%	100%	100%	100%	100%	100%	100%	100%	100%

Notes: C = commercial; R = residential; I = industrial; GIC = government, institution, or community; O = other uses.

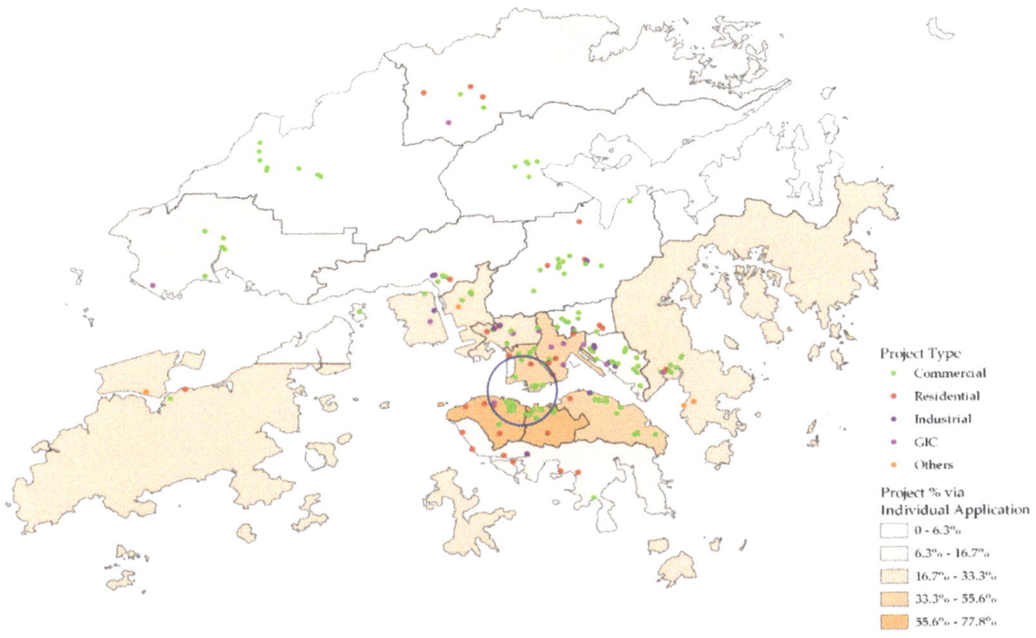

Figure 2. Map showing the project distribution in Hong Kong. Notes: (1) the blue circle indicates the proximate location of Hong Kong's CBD [53]. (2) Boundaries of districts in the figure follow the delineation adopted in 2016 Population By-census [54]. (3) The map was produced using QGIS 3.

4.5. By Completion Years and Building Ages upon Certification

Figure 3 presents two charts, illustrating the completion years of BEAM-certified existing buildings and the information of building ages when the current certifications were obtained. Overall, the building ages of existing buildings when eco-certified are rather diverse though the certification pathway, exerting no significant impact on the building age group distribution (t-statistic = 0.62; $p > 0.1$). Moreover, 75% of the certifications were granted to the buildings between 11 and 40 years of age. A significant difference between commercial and residential projects can be identified by comparing the building age groups of the two sets of projects (t-statistic = 3.88; $p < 0.01$). That is, eco-certified residential buildings tend to be "younger" than their commercial counterparts. Moreover, 73% of BEAM EB certifications were issued to residential projects under 20 years old. Particularly, 44% of certified residential buildings were completed less than 10 years before certification. On the contrary, only 11% of commercial buildings were less than 10 years when awarded the BEAM certification. Over half (53.2%) of the eco-certified existing buildings were commercial buildings between 21 and 40 years of age. Existing buildings built before 1970 were the HK-BEAM-certified existing Grade-A office with longstanding sustainable building maintenance (as discussed in Section 4.2) or renovated in the 2010s. To conclude, comparing the BEAM-certified residential buildings, more "old" commercial buildings are certified under BEAM Plus existing buildings. Thus, building age may be a potential barrier to sustainable building retrofits of residential buildings in Hong Kong.

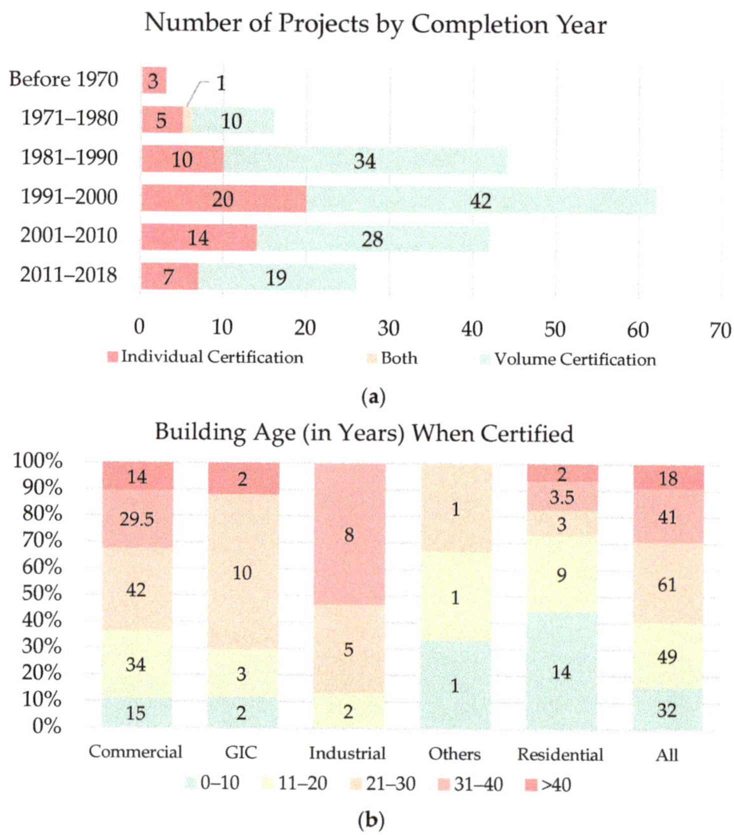

Figure 3. (**a**) Project by completion year; (**b**) building age when certified by project types.

Generally speaking, more architectural and structural constraints may be encountered in the retrofitting projects of the "older" buildings. Higher costs are usually incurred in the building energy retrofits of the old buildings. Furthermore, sufficient evidence shows that the governance of old buildings (such as the old Chinese tenements) is notoriously complicated, particularly when these buildings are so-called the "three-nil buildings" (i.e., buildings not managed by any owners' corporation, residential organization, or external property management agent) [55]. This phenomenon well echoes the findings from the data analysis by completion years and building ages upon certification.

5. Discussion

Drawing on the findings above, disparities have been found in the eco-certification of the existing buildings in Hong Kong. In this section, these disparities will be discussed from various perspectives, particularly the transaction cost theory and agency theory.

5.1. Certification of Multi-Owned Properties Impeded by Institutional Settings

For buildings in multiple ownership, the decision to apply for eco-certification or to undergo building energy retrofits necessitates collective actions among the co-owners. Co-ownership is often regarded as a barrier to the implementation of building energy retrofits [56,57]. The difficulty to initiate eco-certification and retrofits increases with the number of co-owners involved. This echoes the classic Olsonian view that collective actions are less likely to succeed when the group size increases [36]. From a neo-institutional economics perspective, the transaction costs incurred during the negotiation and coordination could be prohibitively high when a large group of co-owners is involved. The transaction costs for initiating eco-certification applications and/or retrofit projects are lower for single-owned projects, which explains why over 70% of the certified EB projects are held in single ownership.

Besides, most BEAM-certified residential projects are of smaller scales. This also goes along with the transaction cost perspective above. The larger is the project scale, the more will be the interested parties (i.e., co-owners) involved. Higher transaction costs will then impede collective actions to partake in eco-certification. Although there are a few large-scale projects in multiple ownership certified under BEAM Plus, these certifications were initiated by the property management companies rather than the co-owners themselves. The agent-led applications for eco-certification can be explained by lower transaction costs incurred in the coordination and lobbying processes. Most other cases of multi-owned residential properties of smaller scales were certified under the selective scheme ("management" aspect) only. The applications were made through volume certifications by the property management companies who managed large portfolios of residential properties throughout the territory. This understanding that greening existing multi-owned properties is more challenging than single-owned properties echoes many previous west studies [58].

The government intends to regard larger housing developments as more resourceful so less subsidization is provided to the large-scale housing developments to initiate building improvement projects. However, the findings of the current research may suggest that in the light of high transaction costs, which impede collective actions, more subsidization should be institutionalised to incentivise co-owners of large-scale housing developments to participate in building eco-certification and retrofit projects.

5.2. Agency Problems of Eco-Certification

As discussed above, property management companies may initiate the eco-certification exercises themselves. They can obtain different "selective benefits" by choosing to participate in the eco-certification. First, the BEAM certificates can showcase their CSR initiatives for fulfilling the ESG requirements. Second, the BEAM certifications obtained by the property management companies in selected projects can serve as marketing tools for promoting the companies. Third, large property management companies have many projects in their management portfolios so they can enjoy discounts in application fees through volume certification. Fourth, as shown in Table 4, some property management companies and/or their parent groups are patron members of the HKGBC who administrates the BEAM Plus scheme. Their participation in the eco-certification exercise can demonstrate their genuine supports to the council and the scheme.

On the other hand, private sector projects tend to get less superior BEAM certifications than the projects owned by the public sector or NGOs. The private sector projects have an average score significantly higher than their non-private sector counterparts (t-statistics = 5.36; $p < 0.01$). Besides, most of the private sector projects are eco-certified under the selective scheme ("management" aspect) only. These findings may indicate that property management agents are not so willing to pay efforts in achieving real energy savings. It is because their managers' remunerations (or profit margins) are set as a certain percentage (usually 10–15%) of the total operating expenses of the building (including utility charges for the common areas and facilities) [59]. In Hong Kong, electricity charge comprises a very large proportion of the expenditure in daily building management [60]. There is a strong incentive for the property management agents to keep the electricity charges high in order to maximise their managers' remunerations. Agency problems exist between the property management agents and their clients (i.e., building owners). In the lack of check and control mechanisms, the agents strive to maximise their own profits at their clients' expenses [61,62]. The agency cost of multi-owned property management is higher when more 'decision power' is dedicated to a property management agent [63]. The high agency cost, intertwined with the high transaction costs of co-owner-led certification, impedes the eco-certification of existing multi-owned buildings or developments.

Moreover, the findings of the current study may suggest that the major developers or sizeable landlords in Hong Kong are more active in eco-certification of their existing properties because of the lower agency costs incurred in the decision-making and execution of eco-certification. In many BEAM-certified EB projects, the property management agents and the developers, landlords, or building owners have close relationships (or they are in the same groups). The parties share compatible or aligned goals, so the agency costs of eco-certification or energy retrofit are comparatively lower.

One of the means to solve the agency problems is to alter the remuneration mechanism for property management agents. Instead of using a cost-plus-margin approach, a fixed amount of service fee for remunerating a property management agent can reduce its disincentive to initiate building energy retrofits. The agent can be further incentivised to retrofit with a bonus contingent on savings in energy consumption or other aspects of environmental performance improvement.

5.3. Greater Drives for Certification of Privately Owned Commercial Properties

Apart from the transaction cost theory and agency theory, we attempt to draw insights in the decision making for eco-certification of existing buildings from other perspectives. In Table 3, one can see that approximately 70% of the certified existing building projects are commercial properties. Residential properties account for only 15% of the certified projects. Apparently, incentives for going green are more significant for commercial properties compared with other building uses. There are several reasons behind the unevenness across building uses. First, many commercial properties, such as offices and retail properties in Hong Kong are for leasing. It has been widely documented that green credentials help landlords attract tenants, particularly institutional tenants, to rent their properties [64,65]. Large companies, especially U.S.-based ones, are very committed to lease eco-certified properties [66,67]. Moreover, rental premiums brought about by the building eco-certifications are quite evident in the commercial property sector [68–70]. Moreover, landlords of single-owned commercial properties may be rewarded as they take the certification as CSR evidence. These drives seem to outweigh the disincentives created by the oft-mentioned dilemma of split incentives for landlords to have their properties go green [71–73].

5.4. More Eco-Certifications with Building Energy Retrofits in the Non-Private Sector

The project type (or building use) is highly related to the sector and ownership type. For instance, GIC buildings, in most circumstances, are single-owned by the government departments (public sector) or NGOs. The non-private organizations tend to have their existing building projects accredited under the comprehensive scheme or selective schemes for various aspects rather than merely "Management" aspects. The non-private sector is more willing to undertake energy retrofits to their buildings, going beyond simply taking sustainable building management practices. Apart from the agency perspective discussed above, the pattern can be explained by the motive of the public sector and NGOs. The non-private organizations would like their BEAM-certified projects to be "demonstration projects" to showcase the applications of new technologies and construction practices, to achieve sustainable building. It is important for the diffusion of such technologies and practices to the whole building sector in Hong Kong.

For the public-owned BEAM-certified buildings, they share at least one of the following features: (i) the building serves as a departmental headquarters; (ii) the building is geographically located in Kowloon East under the environmentally sustainable second CBD agenda (Energizing Kowloon East), and (iii) the building has been recently allocated public funding for refurbishment or infrastructure upgrading. The public sector projects can undergo building energy retrofits with sufficient financial resources being granted. Payback is not a necessary consideration for these public sector projects. Therefore, fewer barriers are expected in public sector projects than in private sector buildings.

Of three EB projects owned and managed by NGOs, two are headquarters buildings, one is for a business sustainability organization, and one is for a charity group. Specifically, the two buildings are single-owned and perform as role models for other organizations to follow in retrofitting their existing premises. The remaining one belongs to the Hong Kong Housing Society, a local public housing agency established in a non-governmental institutional setting. It is an old public rental housing estate renovated with the incorporation of green roofs, the use of environmentally friendly and energy-saving materials, and the introduction of an environmental and energy management system.

On the other hand, private enterprises may just target the signalling effects of the BEAM certification or take the certifications as evidence for ESG reporting. They are inclined to pick the most cost-effective options to fulfil the ESG requirements. Therefore, we can see many private enterprises applied for BEAM certification under the selective scheme for the "management" aspect only because this route necessitates minimal financial inputs. Besides, a high proportion of private enterprises opt for certification through the volume certification pathway, which offers discounts in application fees. To further promote energy retrofits among existing buildings in the city, the Hong Kong government should offer more incentives to the building owners. Apart from the current policy that capital expenditure on the installation of environmentally friendly machinery and equipment can be tax-deductible—the government may consider subsidizing building owners to apply for eco-certification under the comprehensive scheme. The rationale is that existing buildings usually need to be retrofitted first to get eco-certified under the comprehensive scheme.

6. Conclusions

The research findings demonstrate the imbalance in the popularity of eco-certification among different types of existing buildings in Hong Kong. They offer insights into the areas in which the promotion of retrofits is needed. Furthermore, this study opens up a new avenue for broadening the research area of eco-certification of existing buildings. The current research unveils that there are fewer eco-certified existing buildings in multiple ownership and with larger scales. Such finding echoes the transaction cost economics and the classic Olsonian view of collective actions. Besides, agency problems are found to occur in the eco-certification of existing buildings. Property management agents tend to obtain less superior classes of eco-label or certification. This phenomenon reflects that the property management agents are reluctant to initiate energy retrofit projects with the existing buildings because real energy savings are in contradiction with their profit maximization initiatives. Based on the empirical findings, there is a need to rethink the subsidization strategy and redesign the incentive structures for third-party property management services in order to stimulate more existing buildings to be eco-certified.

While we discussed the principal–agent relationship between building owners and property management agents in the current article, agency problems in building energy retrofits also exist between the government and building owners [6]. How these agency problems shape the landscape of eco-certification remains unanswered. Furthermore, there could be some cases where the building owners may want to retrofit their properties but not pursue third-party certification of the projects [74]. There is a possible gap between energy retrofits and eco-certification, particularly for non-investor or non-corporate building owners. Thus, further investigations targeting these issues are warranted.

Author Contributions: Conceptualization, Y.Y. and H.H.; methodology, Y.Y.; formal analysis, Y.Y. and K.C.Y.; data curation, K.C.Y.; writing—original draft preparation, Y.Y., H.H., and K.C.Y.; writing—review and editing, Y.Y., H.H., and Q.K.Q.; visualization, K.C.Y.; supervision, Y.Y. and H.H. All authors have read and agreed to the published version of the manuscript.

Funding: This research received no external funding.

Institutional Review Board Statement: Not applicable.

Informed Consent Statement: Not applicable.

Data Availability Statement: The data presented in this study come from the BEAM Plus Project Directory & Statistics, which are available on the Hong Kong Green Building Council's website (https://www.hkgbc.org.hk/eng/beam-plus/beam-plus-dir-stat/BEAMPlusDirectory.jsp, accessed on 21 May 2021).

Conflicts of Interest: The authors declare no conflict of interest.

References

1. United Nations Environment Programme. *2020 Global Status Report for Buildings and Construction: Towards a Zero-Emission, Efficient and Resilient Buildings and Construction Sector*; United Nations Environment Programme: Nairobi, Kenya, 2020.
2. European Commission. In Focus: Energy Efficiency in Buildings. Available online: https://ec.europa.eu/info/news/focus-energy-efficiency-buildings-2020-feb-17_en (accessed on 25 May 2021).
3. International Energy Agency. Policy Database. Available online: https://www.iea.org/policies (accessed on 25 May 2021).
4. Chu, O.K.; Cheung, R.Y. A Comparison on Two Certification Systems: Leadership in Energy and Environmental Design (LEED) And Building Environmental Assessment Method (BEAM) On Green Building in Hong Kong. In Track 3: Advancing SBE Assessments. In Proceedings of the World Sustainable Built Environment Conference 2017, Hong Kong, China, 9 June 2017.
5. Environment Bureau, Development Bureau, Transport and Housing Bureau. *Energy Saving Plan for Hong Kong's Built Environment 2015–2025+*; Hong Kong SAR Government: Hong Kong, China, 2015.
6. Liang, X.; Yu, T.; Hong, J.; Shen, G.Q. Making incentive policies more effective: An agent-based model for energy-efficiency retrofit in China. *Energy Policy* **2019**, *126*, 177–189. [CrossRef]
7. Darko, A.; Chan, A.P. Review of barriers to green building adoption. *Sustain. Dev.* **2017**, *25*, 167–179. [CrossRef]
8. Qian, Q.K.; Fan, K.; Chan, E.H.W. Regulatory incentives for green buildings: Gross floor area concessions. *Build. Res. Inf.* **2016**, *44*, 675–693. [CrossRef]
9. Fan, K.; Chan, E.H.W.; Qian, Q.K. Transaction costs (TCs) in green building (GB) incentive schemes: Gross floor area (GFA) concession scheme in Hong Kong. *Energy Policy* **2018**, *119*, 563–573. [CrossRef]
10. Friedman, C.; Becker, N.; Erell, E. Energy retrofit of residential building envelopes in Israel: A cost-benefit analysis. *Energy* **2014**, *77*, 183–193. [CrossRef]
11. Fan, K.; Chan, E.H.W.; Chau, C.K. Costs and benefits of implementing green building economic incentives: Case study of a gross floor area concession scheme in Hong Kong. *Sustainability* **2018**, *10*, 2814. [CrossRef]
12. Liu, Y.; Liu, T.; Ye, S.; Liu, Y. Cost-benefit analysis for Energy Efficiency Retrofit of existing buildings: A case study in China. *J. Clean. Prod.* **2018**, *177*, 493–506. [CrossRef]
13. Olaussen, J.O.; Oust, A.; Solstad, J.T. Energy performance certificates—Informing the informed or the indifferent? *Energy Policy* **2017**, *111*, 246–254. [CrossRef]
14. Olaussen, J.O.; Oust, A.; Solstad, J.T.; Kristiansen, L. Energy performance certificates: The role of the energy price. *Energies* **2019**, *12*, 3563. [CrossRef]
15. Shaikh, P.H.; Shaikh, F.; Sahito, A.A.; Uqaili, M.A.; Umrani, Z. An Overview of the Challenges for Cost-Effective and Energy-Efficient Retrofits of the Existing Building Stock. In *Cost-Effective Energy Efficient Building Retrofitting: Materials, Technologies, Optimization and Case Studies*; Pacheco-Torgal, F., Granqvist, C., Jelle, B.P., Vanoli, G.P., Bianco, N., Kurnitski, J., Eds.; Woodhead Publishing: Cambridge, UK, 2017; pp. 257–278.
16. Luther, M.B.; Rajagopalan, P. Defining and developing an energy retrofitting approach. *J. Green Build.* **2014**, *9*, 151–162. [CrossRef]
17. Weiss, J.; Dunkelberg, E.; Vogelpohl, T. Improving policy instruments to better tap into homeowner refurbishment potential: Lessons learned from a case study in Germany. *Energy Policy* **2012**, *44*, 406–415. [CrossRef]
18. Caputo, P.; Pasetti, G. Overcoming the inertia of building energy retrofit at municipal level: The Italian challenge. *Sustain. Cities Soc.* **2015**, *15*, 120–134. [CrossRef]

19. Hong, T.; Piette, M.A.; Chen, Y.; Lee, S.H.; Taylor-Lange, S.C.; Zhang, R.; Sun, K.; Price, P. Commercial building energy saver: An energy retrofit analysis toolkit. *Appl. Energy* **2015**, *159*, 298–309. [CrossRef]
20. Hou, J.; Liu, Y.; Wu, Y.; Zhou, N.; Feng, W. Comparative study of commercial building energy-efficiency retrofit policies in four pilot cities in China. *Energy Policy* **2016**, *88*, 204–215. [CrossRef]
21. Building Performance Institute Europe. *Europe's Buildings under the Microscope—A Country-by-Country Review of the Energy Performance of Buildings*; Buildings Performance Institute Europe: Brussel, Belgium, 2011.
22. Amecke, H.; Deason, J.; Hobbs, A.; Novikova, A.; Xiu, Y.; Zhang, S. *Buildings Energy Efficiency in China, Germany, and the United States*; Climate Policy Initiative: San Francisco, CA, USA, 2013.
23. Abdallah, M.; El-Rayes, K.; Liu, L. Minimizing upgrade cost to achieve LEED certification for existing buildings. *J. Constr. Eng. Manag.* **2016**, *142*, 04015073. [CrossRef]
24. Aktas, B.; Ozorhon, B. Green building certification process of existing buildings in developing countries: Cases from Turkey. *J. Manag. Eng.* **2015**, *31*, 05015002. [CrossRef]
25. Ma, Z.; Wang, S. Building energy research in Hong Kong: A review. *Renew. Sustain. Energy Rev.* **2009**, *13*, 1870–1883. [CrossRef]
26. Electrical and Mechanical Services Department (EMSD). Hong Kong Energy Efficiency Registration Scheme for Buildings (2018 Edition). Available online: https://www.emsd.gov.hk/filemanager/en/content_723/hkeersb2018_en.pdf (accessed on 18 May 2021).
27. Electrical and Mechanical Services Department (EMSD). Hong Kong Energy Efficiency Registration Scheme for Buildings. Available online: https://www.emsd.gov.hk/en/energy_efficiency/energy_efficiency_registration_scheme_for_building/index.html (accessed on 18 May 2021).
28. Hong Kong Green Building Council (HKGBC). BEAM Plus—Introduction. Available online: https://www.hkgbc.org.hk/eng/beam-plus/introduction/index.jsp (accessed on 5 July 2021).
29. Hong Kong Green Building Council, BEAM Society Limited. *BEAM Plus Existing Buildings Version 2.0 (2016.03): Comprehensive Scheme*; BEAM Society Limited: Hong Kong, China, 2016.
30. Hong Kong Green Building Council, BEAM Society Limited. *BEAM Plus Existing Buildings Version 2.0 (2016.03): Selective Scheme*; BEAM Society Limited: Hong Kong, China, 2016.
31. Coase, R.H. The Nature of the Firm. *Economica* **1937**, *4*, 386–405. [CrossRef]
32. Coase, R.H. The Problem of Social Cost. *J. Law Econ.* **1960**, *3*, 1–44. [CrossRef]
33. Gordon, R.L. Regulation and Economic Analysis: A Critique Over Two Centuries. Kluwer Academic Publishers: Dordrecht, The Netherlands, 1994.
34. Williamson, O.E. Transaction-cost economics: The governance of control relations. *J. Law Econ.* **1979**, *22*, 233–261. [CrossRef]
35. Williamson, O.E. The economics of Organization: The transaction cost approach. *Am. J. Sociol.* **1981**, *87*, 548–577. [CrossRef]
36. Olson, M. *The Logic of Collective Action: Public Goods and the Theory of Groups*; Harvard University Press: Cambridge, MA, USA, 1965.
37. Yau, Y. Multicriteria decision making for homeowners' participation in building maintenance. *J. Urban Plan. Dev.* **2011**, *138*, 110–120. [CrossRef]
38. Jia, L.; Qian, Q.K.; Meijer, F.; Visscher, H. Exploring key risks of energy retrofit of residential buildings in China with transaction cost considerations. *J. Clean. Prod.* **2021**, *293*, 126099. [CrossRef]
39. Walker, A.; Chau, K.W. The relationship between construction project management theory and transaction cost economics. *Eng. Constr. Archit. Manag.* **1999**, *6*, 166–176. [CrossRef]
40. Ahn, Y.H.; Pearce, A.R. Green construction: Contractor experiences, expectations, and perceptions. *J. Green Build.* **2007**, *2*, 106–122. [CrossRef]
41. Meacham, B.J. Accommodating innovation in building regulation: Lessons and challenges. *Build. Res. Inf.* **2010**, *38*, 686–698. [CrossRef]
42. Chai, K.; Yeo, C. Overcoming energy efficiency barriers through systems approach—A conceptual framework. *Energy Policy* **2012**, *46*, 460–472. [CrossRef]
43. Qi, G.Y.; Shen, L.Y.; Zeng, S.X.; Jorge, O.J. The drivers for contractors' green innovation: An industry perspective. *J. Clean. Prod.* **2010**, *18*, 1358–1365. [CrossRef]
44. Häkkinen, T.; Belloni, K. Barriers and drivers for sustainable building. *Build. Res. Inf.* **2011**, *39*, 239–255. [CrossRef]
45. Jensen, M.; Meckling, W. Theory of the firm: Managerial behaviour, agency costs, and ownership structure. *J. Financ. Econ.* **1976**, *3*, 305–360. [CrossRef]
46. Moore, J.H. Agency costs, technical change and Soviet central planning. *J. Law Econ.* **1981**, *24*, 189–214. [CrossRef]
47. Fama, E.; Jensen, M. Separation of ownership and control. *J. Law Econ.* **1983**, *26*, 301–326. [CrossRef]
48. Rauchhaus, R.W. Principal-agent problems in humanitarian intervention: Moral hazards, adverse selection, and the commitment dilemma. *Int. Stud. Q.* **2008**, *53*, 871–884. [CrossRef]
49. Fayezi, S.; O'Loughlin, A.; Zutshi, A. Agency theory and supply chain management: A structured literature review. *Supply Chain Manag.* **2012**, *17*, 556–570. [CrossRef]
50. Calvo, N.; Calvo, F. Corporate social responsibility and multiple agency theory: A case study of internal stakeholder engagement. *Corp. Soc. Responsib. Environ. Manag.* **2018**, *25*, 1223–1230. [CrossRef]

51. Delbufalo, E. Agency Theory and Sustainability in Global Supply Chain. In *Agency Theory and Sustainability in the Global Supply Chain*; Springer: Cham, Switzerland, 2018; pp. 33–54.
52. Kumbaroğlu, G.; Madlener, R. Evaluation of economically optimal retrofit investment options for energy savings in buildings. *Energy Build.* **2012**, *49*, 327–334. [CrossRef]
53. Development Bureau, Planning Department. *Hong Kong 2030+: Towards a Planning Vision and Strategy Transcending 2030*; Hong Kong SAR Government: Hong Kong, China, 2016; pp. 45–46.
54. ESRI China (HK). Hong Kong 18 Districts. Available online: https://opendata.esrichina.hk/datasets/eea8ff2f12b145f7b33c4eef4f045513_0/explore?location=22.357821%2C114.139117%2C11.58&showTable=true (accessed on 9 June 2021).
55. Audit Commission. *Director of Audit's Report No. 75*; Audit Commission: Hong Kong, China, 2020.
56. Altmann, E. Apartments, co-ownership and sustainability: Implementation barriers for retrofitting the built environment. *J. Environ. Policy Plan.* **2014**, *16*, 437–457. [CrossRef]
57. Bright, S.; Weatherall, D. Framing and mapping the governance barriers to energy upgrades in flats. *J. Environ. Law* **2017**, *29*, 203–229. [CrossRef]
58. Maruejols, L.; Young, D. *Energy Use in Canadian Multi-Family Dwellings*; Canadian Building Energy End-Use Data and Analysis Centre: Edmonton, AB, Canada, 2010.
59. The Hong Kong Institute of Surveyors. *Standard Property Management Agreement*; The Hong Kong Institute of Surveyors: Hong Kong, China, 2013.
60. Property and Facility Management Division of the Hong Kong Institute of Surveyors. *Benchmarking of Management Fees for Office Buildings in Hong Kong*; The Hong Kong Institute of Surveyors: Hong Kong, China, 2010.
61. Yau, Y.; Ho, D.C.W.; Li, R. Benchmarking property management agents' performance in Hong Kong. *Int. J. Dev. Sustain.* **2017**, *6*, 650–666.
62. Gao, L.W.; Ho, D.C.W. Explaining the outcomes of multi-owned housing management: A collective action perspective. *Habitat Int.* **2016**, *57*, 233–241. [CrossRef]
63. Chu, F.-N.; Chang, C.-O. Performance evaluation of condominium management and maintenance modes: From a perspective of principal-agent relationship. *Manag. Rev.* **2013**, *32*, 123–130.
64. Jones Lang Lasalle. *The Impacts of Sustainability on Value: Developing the Business Case for Net Zero Carbon Buildings in Central London*; Jones Lang Lasalle: London, UK, 2020.
65. Matisoff, D.C.; Noonan, D.S.; Mazzolini, A.M. Performance or marketing benefits? The case of LEED certification. *Environ. Sci. Technol.* **2014**, *48*, 2001–2007. [CrossRef]
66. Bower, B.; Boyd, N.; McGoun, E. Greenbacks, green banks, and greenwashing via LEED: Assessing banks' performance in sustainable construction. *Sustain. J. Rec.* **2020**, *13*, 208–217. [CrossRef]
67. Eichholtz, P.; Kok, N.; Quigley, J.M. Why companies rent green: CSR and the role of real estate? *Inst. Bus. Econ. Res.* **2017**, *2009*, W09-004. [CrossRef]
68. Eichholtz, P.; Kok, N.; Quigley, J.M. The economics of green buildings. *Rev. Econ. Stat.* **2013**, *95*, 50–63. [CrossRef]
69. Wadu, M.J.; Chan, H.T. Environmental certification schemes and property values: Evidence from the Hong Kong prime commercial office market. *Int. J. Strateg. Prop. Manag.* **2019**, *23*, 81–95.
70. Li, W.; Fang, G.; Yang, L. The effect of LEED certification on office rental values in China. *Sustain. Energy Technol. Assess.* **2021**, *45*, 101182.
71. Collins, D.; Junghans, A.; Haugen, T. Green leasing in commercial real estate: The drivers and barriers for owners and tenants of sustainable office buildings. *J. Corp. Real Estate* **2018**, *20*, 244–259. [CrossRef]
72. Wrigley, K.; Crawford, R.H. Identifying policy solutions for improving the energy efficiency of rental properties. *Energy Policy* **2017**, *108*, 369–378. [CrossRef]
73. Yang, E.; Guevara-Ramirez, J.S.; Bisson, C. Finding evidence of green leasing in United States government-leased properties. *J. Green Build.* **2020**, *15*, 55–72. [CrossRef]
74. Qiu, Y.L.; Su, X.; Wang, D. Factors influencing commercial buildings to obtain green certificates. *Appl. Econ.* **2017**, *49*, 1937–1949. [CrossRef]

energies

Article

Key Performance Indicators for Evaluation of Commercial Building Retrofits: Shortlisting via an Industry Survey

Man Ying (Annie) Ho [1], Joseph H. K. Lai [1], Huiying (Cynthia) Hou [1] and Dadi Zhang [2,*]

1 Department of Building Environment and Energy Engineering, The Hong Kong Polytechnic University, Hong Kong 999077, China; annieho1010@yahoo.com.hk (M.Y.H.); bejlai@polyu.edu.hk (J.H.K.L.); cynthia.hou@polyu.edu.hk (H.H.)
2 Faculty of Architecture and the Built Environment, Delft University of Technology, 2628 BL Delft, The Netherlands
* Correspondence: D.Zhang-2@tudelf.nl

Abstract: Key performance indicators (KPIs) are quintessentially useful for performance evaluation, but a set of pragmatic KPIs for holistic evaluation of retrofits for commercial buildings is hitherto unavailable. This study was conducted to address this issue. Built upon the findings of a systematic literature review and a focus group meeting in the earlier stages of the study, a questionnaire survey covering 19 KPIs for environmental (embracing energy), economic, health and safety, and users' perspective evaluations of building retrofits was developed. Data of the survey, collected from facility management (FM) practitioners in Hong Kong, underwent a series of statistical analyses, including Kruskal–Wallis H test, Mann–Whitney U test, and Spearman Rank Correlation. The analysis results revealed the levels of importance of KPIs perceived by different groups of FM practitioners and the rankings of KPIs. Based upon these results, eight KPIs were shortlisted, which are energy savings, payback period, investment cost, actual-to-target ratio of the number of statutory orders removed, actual-to-target ratio of the number of accidents reduced, target indoor air temperature, target indoor air quality (IAQ) class, and target workplane illuminance. These KPIs serve as keystones for further development of an analytic evaluation scheme for commercial building retrofit performance assessment. The methodology of this study can also serve as a reference for similar KPI studies in other research domains.

Keywords: facility management; KPI; refurbishment; renovation; retrofit; survey

Citation: Ho, M.Y.; Lai, J.H.K.; Hou, H.; Zhang, D. Key Performance Indicators for Evaluation of Commercial Building Retrofits: Shortlisting via an Industry Survey. *Energies* **2021**, *14*, 7327. https://doi.org/10.3390/en14217327

Academic Editor: F. Pacheco Torgal

Received: 15 September 2021
Accepted: 29 October 2021
Published: 4 November 2021

1. Introduction

Buildings account for 39% of all carbon emissions in the world [1]. In Hong Kong—a city famous for its dense population and buildings—the volume of aged buildings is large and keeps increasing. As retrofitting those existing buildings is a sustainability goal that the international society endeavours to meet, the building industry and the government of Hong Kong have introduced various incentives that motivate building owners or operators to implement retrofits for the premises they own or manage. However, the retrofit rate of existing buildings remains low [2]. The building sector in the city is facing challenges to retrofitting the existing buildings, especially the aged buildings. One of the key challenges is the estimation of the benefits brought by the building retrofits, which relies on scientific evaluation mechanisms to evaluate the building retrofit performance against the economic input [3,4]. In the evaluation process, human decisions or judgements from owners, operators, occupants, etc. are critical elements.

Facility management (FM) practitioners are building professionals who are involved in multiple disciplines of practice to ensure the functionality, comfort, safety, and efficiency of facilities in the built environment. Their knowledge and work experience are gained through intensive interactions with the operations of the existing buildings they manage. Thus, their opinions on the change in the buildings' conditions and the best option for building upgrades should be valid and reliable. Although a profusion of studies has been undertaken to help building owners and operators make decisions on building retrofits, there is still limited research on establishing a set of key performance indicators (KPIs) for the development of an analytic method to evaluate the holistic performance of commercial building retrofits [5]. To address this research gap, this multi-stage study was initiated. In the initial stages, a systematic literature review was conducted to identify indicators that are applicable to performance evaluation for building retrofits, followed by a focus group meeting with FM experts, from which the experts' perceived importance of the indicators were solicited, and the practicability of the indicators in real-world commercial buildings was discussed [6]. The outcome of those preceding stages, i.e., selection of 19 essential KPIs among the 52 applicable performance indicators identified from the literature, was adopted to design a questionnaire survey. Distributed to FM practitioners in Hong Kong, the survey intended to solicit opinions from a large sample of industry professionals on their perceived importance of the 19 KPIs. This paper, reporting on the works undertaken for this survey, is structured as follows: Sections 2 and 3 show the literature review and the data collection method; Sections 4 and 5 present the statistical analyses and results of the analysed findings; Section 6 discusses the implications of this study; Section 7 concludes the paper.

2. Building Retrofit

2.1. Definitions, Scope, and Challenges

Building retrofit, a form of technical intrusion in the systems or structure of a building after its initial construction and occupation [7], can improve the building's performance to optimise energy utilisation and enhance users' occupation experience. In recent years, building retrofit is increasingly referred to as green retrofitting or sustainable retrofitting, which emphasises the environmental benefits of retrofitting work in the built environment. The US Green Building Council (USGBC) defines green retrofitting as '... any upgrade of an existing facility to improve energy and environmental performance, decrease usage of water and enhance existing comfort and quality of interior spaces—All achieved in a manner that provides financial incentives to the investor' [8]. Common building retrofit projects include energy efficiency retrofits (e.g., upgrading of the building envelope) [9], lighting retrofits [9,10], ventilation system retrofits [10], water efficiency retrofits [11], renewable energy installations [9,10], green roof establishment [12], building automation and control system [13], and space utilisation and reconfiguration.

In general, building retrofit has been facing numerous challenges, including but not limited to, optimisation or fulfilment of priorities of stakeholders, time period, capital investment, cost effectiveness, risk analysis, technology availability, government policies, and building energy performance prediction [7]. From a process management perspective, these challenges are influenced by each other in an interactive manner and some of them incur extra costs, which can become liabilities to the building owners or its occupants. In order to support building retrofit decisions, various evaluation methodologies have been developed to provide an estimation of the associated energy consumption and cost, thereby facilitating the building retrofit process design [14]. The manifestation of such challenges varies with the project or building specificities, such as building types (e.g., public building, residential building, and office building), building design (including both system and structural design), building material, and building technologies [15,16].

2.2. Performance Measurement

In the past decade, a considerable number of studies have been conducted in evaluating the performance of building retrofits. Decision making is a prevailing stream of studies in the field of building retrofits. Economic viability is an indicator frequently used for building retrofit performance measurement. Net present value (NPV), internal rate of return (IRR), overall rate of return (ORR), benefit–cost ratio (BCR), discounted payback period (DPP), and simple payback period (SPP) are often used to assess the economic feasibility of a single retrofit measure [15,17–20]. Surveys on users' satisfaction or feedback from stakeholders in the post-occupancy phase were also used to measure building retrofit performances [21,22].

On top of that, the KPI approach is regarded to be one of the most popular and valuable tools for measuring the process or outcome of construction projects. KPIs are a collection of indicators that can comprehensively reflect a project's goals. They help to define the nature, scope, expected quality, and unique characteristics of the projects and can also provide means for measuring the 'progress towards those goals for further learning and improvement' [23]. Energy performance and energy saving are two common KPIs for measuring the financial and environmental benefits of building retrofits [24–28]. As sustainability is one of the key project goals, an increasing volume of studies in the literature have examined KPIs for measuring the level of sustainability in construction and building renovation/retrofit projects. Kylili et al. [23] provided a state-of-the-art review on the KPIs identified for measuring the sustainability of the projects in the built environment, in which they categorised the building performance KPIs into eight groups—namely, economic, environmental, social, technical, time, quality, disputes, and project administration. Al Dakheel et al. [29] conducted a review on features of smart buildings (SBs) and identified 10 KPIs for SBs. The KPIs they identified help to quantify the 'smart features' of SBs and reflect the 'smart capability' of the building. The validity of KPIs affects the overall measurement results; thus, the selection process of KPIs should engage scientific methodologies to ensure their representativeness of the measurement goals [30]. Industry experts' involvement is regarded as one of the reliable approaches for identifying representative and valid KPIs. This approach usually entails three steps: (1) interviews with experts to define the measurement goals and identify KPIs that fit those goals; (2) a survey to collect a wider scale of data from various groups of experts; (3) statistical analyses to confirm and verify the identified KPIs from the previous steps. Xu et al. [31] followed these three steps to identify KPIs for the sustainability of building energy efficiency retrofit (BEER) in hotel buildings in China. Lai et al. [32] used the same approach to investigate KPIs for measuring the performance of hospital facilities management. This study, likewise, adopted this approach to identify and verify KPIs for the evaluation of commercial building retrofits.

2.3. The Role of FM Practitioners

A building retrofit project usually comprises five major phases: project set up and pre-retrofit survey, performance assessment, identification of retrofit options, site implementation and commission, and validation and verification [15]. Completion of a retrofitting project requires a team of building experts to assess the existing building conditions, design the retrofitting strategies, monitor the retrofitting process, and review the project outcome [33]. In this process, FM practitioners deal with daily building management activities at the operational level and are involved in developing cost-effective plans to support built asset management at the strategic level. For example, FM managers are engaged in a company's corporate social responsibility strategy development through evaluating the facility performance of the property portfolio. They are responsible for providing advice to top management on green certification decisions and participate in obtaining green certification. Building retrofit is an inevitable activity that leads to green certification for existing buildings.

Responsible for managing both buildings and the relevant stakeholders (e.g., owners, occupants, tenants), FM practitioners have to communicate with retrofitting decision makers and facility users, modify facilities, upgrade systems for energy use, and develop mechanisms for measuring energy consumption, monitoring energy use process, and assessing energy performance [34,35]. Thus, FM practitioners play critical roles in supporting decision making on building retrofits, and their opinions on KPIs for building retrofit performance are useful.

3. Data Collection

In the preliminary stages of this study, 52 performance indicators for building retrofit performance assessment were identified through the Preferred Reporting Items for Systematic Reviews and Meta-Analyses (PRISMA) process and these indicators were grouped into four categories ('economic', 'environmental', 'health and safety', and 'users' perspective') with their detailed meanings provided [6]. With these indicators identified and in order to establish a method that can evaluate the holistic performance of commercial building retrofits, a focus group study was then conducted. Grounded upon the deliberations and opinions of the focus group study, 19 KPIs were selected as useful for building retrofit performance assessment (Table 1). Since using 19 KPIs entails a considerable effort to collect the associated empirical FM data, and the process for calculating this large number of KPIs is time consuming [36–38], a questionnaire survey was designed to shortlist indicators that are useful for reflecting the performance of commercial building retrofits.

Table 1. Performance indicators selected from the preliminary stages of the study.

No.	Indicator	Aspect (No. of Indicators)
1	Energy savings (%)	Environmental (5)
2	Normalised energy savings (kWh/m^2 year)	
3	Electricity consumption saving per year (kWh/year)	
4	Energy payback period (year)	
5	Target green building label	
6	Payback period (year)	Economic (7)
7	Return on investment (%)	
8	Internal rate of return (%)	
9	Investment cost (USD)	
10	Normalised investment cost (USD/m^2)	
11	Life cycle cost (USD)	
12	Increase of building value (%)	
13	Ratio of actual to target no. of statutory orders removed (%)	Health and Safety (2)
14	Ratio of actual to target no. of accidents per year reduced (%)	
15	Target indoor air temperature (°C)	Users' perspective (5)
16	Δ Indoor carbon dioxide levels or harmful substances (ppm)	
17	Target IAQ class (good/excellent level)	
18	Target workplane illuminance (lux)	
19	Target equivalent continuous weighted sound pressure level (dBA)	

The questionnaire consists of three parts. Part 1 collects respondents' personal information, including gender, years of work experience, job level, nature of their organisation, type of employer, and their academic qualification. These pieces of information served to reflect the backgrounds of the respondents, allowing inter-group comparisons to be made when analysing the survey findings. Part 2 solicits the importance ratings of the 19 KPIs on a five-point scale (1: very low; 2: low; 3: moderate; 4: high; and 5: very high). Part 3 asks the participants to suggest any other KPIs they consider important and any other comments they have based on their experience.

Pilot tests, with the participation of five FM experts, were conducted on the questionnaire. These tests helped to detect and eliminate any potential error or misunderstanding of the questions in the survey. Feedbacks from the tests were taken to finalise the questionnaire before its official distribution. The industry-wide online survey was officially launched in two ways: snowballing and mass email. Using a snowballing approach, FM professionals who participated in the preceding focus group study [6] and pilot tests were invited to complete the survey and also distribute it to their colleagues. As regards the second approach, mass email, a hyperlink to the survey was emailed to the members of the Building Services Operation and Maintenance Executives Society (BSOMES)—the leading professional body in Hong Kong specialised in technical FM works embracing building retrofits. In order to increase the level of representativeness of the samples, FM practitioners with different organisational natures (government, non-governmental organisation (NGO), and private company) and types (e.g., owner/developer, management company, and contractor) working at different levels (strategic (e.g., director, chief engineer), tactical (e.g., manager, engineer)) were invited to participate in the survey. The demographic details of the survey respondents are shown in Table 2.

Table 2. Demographic details of respondents.

Characteristic	Subgroup	Number	Percentage
Gender	Male	104	83.9%
	Female	20	16.1%
Work experience	≤5 years	22	17.9%
	>5 to <20 years	21	17.1%
	20 to <30 years	39	31.7%
	≥30 years	42	33.9%
Nature of organisation	Government	11	8.9%
	NGO	23	18.6%
	Private company	90	72.6%
Type of employer	Owner/developer	43	34.7%
	Management company	44	35.5%
	Contractor	18	14.5%
	Others	19	15.3%
Job level	Strategic	38	30.7%
	Tactical	86	69.4%
Academic qualification	Associate degree/diploma/certificate	7	5.6%
	Bachelor degree	32	25.8%
	Master degree	81	65.3%
	Doctorate degree	2	1.6%
	Others	2	1.6%

A total of 164 responses to the survey were received. To ensure data quality, the responses were screened manually, and those with incomplete information provided were discarded. This resulted in having 124 responses qualified for the subsequent data analysis. Among these responses, 83.9% were from males. The majority of the respondents were highly experienced; most were employed by private companies. The proportions of those working for owners/developers and management companies were comparable, while those working for contractors amounted to 14.5%. When comparing the strategic and tactical groups, the latter prevails. More than three-quarters of the participants have worked on office buildings; nearly half have worked on retail premises. The respondents were well educated, with most of them possessing a degree at the bachelor level or above.

4. Statistical Analysis

The data collected were analysed using the SPSS version 26.0 software. To investigate any differences between different groups of the responses, the respondents were stratified into six groups. Each of these groups was further categorised into subgroups ('n' denotes the number of samples), as shown below:

- G1: Gender: male (G1a; $n = 104$) and female (G1b; $n = 20$);
- G2: FM/operation and maintenance (O&M) work experience: $\leq$5 years (G2a; $n = 22$), 5 years to <20 years (G2b; $n = 21$), 20 to <30 years (G2c; $n = 39$), and $\geq$30 years (G2d; $n = 42$);
- G3: Nature of organisation that the respondents worked for: government (G3a; $n = 11$), public (G3b; $n = 23$) and private (G3c; $n = 90$);
- G4: Type of employer: owners/developers (G4a; $n = 43$), management companies (G4b; $n = 44$), contractors (G4c; $n = 18$) and others (G4d; $n = 19$);
- G5: Job level: strategic (G5a; $n = 38$) and tactical (G5b; $n = 86$);
- G6: Academic qualification: subdegree (associate degrees/diplomas/certificates), bachelor (G6a; $n = 41$), and postgraduate (master degrees or doctorate degrees) (G6b; $n = 83$).

First, group analyses were conducted using Kruskal–Wallis H test (H) and Mann–Whitney U test (U) to analyse whether the respondents perceived the importance levels of the KPIs differently. H test, a non-parametric test that compares more than two independent or unrelated samples [39], was applied to make comparisons between groups G2, G3, and G4. For each of the comparisons, a null hypothesis (H_o) and an alternative hypothesis (H_a) were set (H_o: there is no tendency for ranks of groups of response to rank systematically higher or lower than those of the others; H_a: there is a tendency for ranks of groups of response to rank systematically higher or lower for at least one of the groups). For testing these hypotheses, the Kruskal–Wallis H-test statistic was determined by Equation (1).

$$H = \frac{12}{N(N+1)} \sum_{i=1}^{k} \frac{R_i^2}{n_i} - 3(N+1) \tag{1}$$

where N is the number of values from all combined samples, R_i is the sum of the ranks from a particular sample, and n_i is the number of values from the corresponding rank sum. The α value for the H test was set as 0.05. As H test does not identify where and the degree of the differences occurred, the U test, as a type of post hoc test, was used to analyse any significant differences between the sample pairs. Following the Bonferroni procedure which helps to compensate type I error, the adjusted α, α_B, was used in the U test to determine any significant difference between samples if H test was found to be significant (e.g., for G2, k = 4) [39]. The adjusted α from the Bonferroni procedure was shown in Equation (2).

$$\alpha_B = \frac{\alpha}{k} \tag{2}$$

where α_B is the adjusted level of risk, α is the original level of risk, and k is the number of comparisons.

For the U test, it is a non-parametric test that can be used to compare two unrelated or independent samples [39]. Therefore, it is suitable for use in inter-group comparisons between G1a and G1b, G5a and G5b, and G6a and G6b. The α value for the U test was set as 0.05. For each of the comparisons, a null hypothesis (H_o) and an alternative hypothesis (H_a) were set (H_o: the mean ranks of the groups are the same; H_a: the rank of one group of responses is systematically higher (or lower) than the other). Accordingly, the Mann–

Whitney U-test statistic for each of the two samples was determined by Equation (3), and the smaller of the two U statistics was obtained.

$$U_i = n_1 n_2 + \frac{n_i(n_i + 1)}{2} - \sum R_i \quad (3)$$

where U_i is the test statistic for the sample of interest, n_i is the number of values from the sample of interest, n_1 is the number of values from the first sample, n_2 is the number of values from the second sample, and ΣR_i is the sum of the ranks from the sample of interest. The mean and z-score for the Mann–Whitney U test for large samples were found by Equations (4) and (5).

$$\overline{x_U} = \frac{n_1 n_2}{2} \quad (4)$$

$$z = \frac{U_i - \overline{x_U}}{S_U} \quad (5)$$

where $\overline{x_U}$ is the mean, S_U is the standard deviation, and z is the z-score for a normal approximation of the data.

Second, Spearman's rank correlation was applied to examine any significant difference in the KPIs rankings between pairs of the respondent subgroups. Spearman's rank correlation coefficient (r_s) was calculated by Equation (6) [34]:

$$r_s = 1 - \frac{6 \sum D_i^2}{n(n^2 - 1)} \quad (6)$$

where n is the number of observations, and D_i is the difference between ranks obtained from each pair of responses. For the value of r_s, '+1' represents perfect agreement between the rankings; '0' represents no association between the rankings; '−1' represents perfect disagreement between the rankings.

Finally, a mean score was calculated for each of the rated KPIs, based on which the overall ranking of the KPIs was determined to facilitate shortlisting the most essential KPIs.

5. Results

5.1. Perceived Importance Levels of KPIs

Referring to the results of the H and U tests shown in Figures 1–12 (see Appendix A), six significant observations are worth noting. First, significant difference (U = 701, $p < 0.05$) was found between male (mean rank = 59.24) and female (mean rank = 79.45) for KPI-11 (life cycle cost (USD)). It means that the female and male FM practitioners had different perceptions of the importance of life cycle cost for evaluating building retrofit performance. This finding echoes the argument of Rodríguez et al. [40] regarding managerial style: men and women have different managerial styles.

Second, significant difference (H = 10.538, $p < 0.05$) was found between respondents with different FM/O&M work experience for KPI-2 (normalised energy savings (kWh/m^2 year)). The results show that (U = 249, $p < 0.0125$) respondents with less work experience (≤5 years; mean rank = 39.18) considered KPI-2 as more important than the experienced practitioners (experience between 20 to 30 years, mean rank = 26.38) did, and similar findings were found between freshmen (≤5 years; mean rank = 41.59) and veterans (≥30 years; mean rank = 27.74) in ranking KPI-2 (U = 262, $p < 0.0125$). This may be because the experienced practitioners were aware that after years of building occupation with energy retrofits already undertaken, the room for further energy saving is limited. Yet, no major disagreement was found between the various respondent groups (with different work experiences) on KPI-1 'energy savings (%)'. According to Miller and Higgins [41], 'percentage better and percentage saved' was mostly referenced in environmental performance evaluation studies [41].

Third, a significant difference (H = 8.726, $p < 0.05$) was found between freshmen and non-freshman (>5 years) for KPI-13 (ratio of actual to target no. of statutory orders removed (%)). The results (U = 125.5, $p < 0.0125$) show that respondents with more work experience (mean rank = 27.02) considered KPI-13 as more important than the freshmen (mean rank = 17.20). The possible reason could be the freshmen may have relatively less work experience and may not have come across any retrofit projects with the requirement in statutory orders removal. Therefore, the freshmen were less concerned about this KPI.

The other three significant differences were found between respondents at the tactical level and strategic level for KPI-17 (target IAQ class; U = 1234, $p < 0.05$); KPI-18 (target workplane illuminance (lux); U = 1159, $p < 0.05$); KPI 19 (target indoor equivalent continuous weighted sound pressure level (dBA); U = 1239, $p < 0.05$). Respondents at the tactical level perceived the three KPIs (mean rank of KPI-17 = 67.15; mean rank of KPI-18 = 68.02; and mean rank of KPI-19 = 67.09) as more important than those at the strategic level did (mean rank of KPI-17 = 51.97; mean rank of KPI-18 = 50.00; mean rank of KPI-19 = 52.11). The reason could be that FM practitioners at the tactical level have to handle and resolve complaints from users about IAQ, workplane illuminance, noise, etc. before such problems escalate to the strategic level. Hence, FM practitioners at the tactical level, when compared with the strategic counterpart, are more concerned about the KPIs in the users' perspective aspect.

5.2. Correlation between Rankings of KPIs

Table 3 displays the values of the Spearman rank correlation coefficients between groups, while the detailed ranking results for different respondent groups are illustrated in Appendix A (Tables 1–3). In general, no significant disagreement was found between the rankings of the KPIs pertinent to the various groups. A significant positive correlation (at the 0.01 level) existed in the rankings between some of the respondent groups, including female vs. male; experience (>5 to <20 years) vs. experience (20 to 30 years); experience (>5 to <20 years) vs. experience (>30 years); owner/developer vs. management company; and sub-degree or undergraduate degree vs. postgraduate degree. Significant positive correlations (at the 0.05 level) existed in the rankings between some of the respondent groups, including experience (20 to 30 years) vs. experience (>30 years); owner/developer vs. others; management company vs. contractors; contractors vs. others; strategic level vs. tactical level.

Table 3. Spearman rank correlations between groups.

Group	Subgroups in Comparison		r_s	p-Value
G1: Gender	Male	Female	0.591 **	0.008
G2: Work experience	(≤5 years)	(>5 to <20 years)	0.396	0.093
		(20 to 30 years)	0.350	0.142
		(>30 years)	0.120	0.624
	(>5 to <20 years)	(20 to 30 years)	0.661 **	0.002
		(>30 years)	0.536 **	0.018
	(20 to 30 years)	(>30 years)	0.505 *	0.027
G3: Nature of organisation	Government	NGO	0.190	0.435
		Private	0.112	0.647
	NGO	Private	0.406	0.085

Table 3. *Cont.*

Group	Subgroups in Comparison		r_s	*p*-Value
G4: Type of employer	Owner/developer	Management company	0.859 **	0.000
		Contractor	0.380	0.108
		Others	0.470 *	0.042
	Management company	Contractor	0.470 *	0.042
		Others	0.370	0.119
	Contractor	Others	0.516 *	0.024
G5: Job level	Strategic level	Tactical level	0.475 *	0.040
G6: Academic qualification	Subdegree orundergraduate degree	Postgraduate degree	0.654 **	0.002

** correlation is significant at the 0.01 level (two-tailed); * correlation is significant at the 0.05 level (two-tailed).

5.3. Importance Levels and Ranks of KPIs

Based on all the valid responses, a mean rating was calculated for each of the KPIs, and the calculation results in Table 4 show that the ratings ranged from 3.14 to 3.76.

Table 4. Mean importance ratings and ranks of KPIs.

	KPI	Mean	Rank	Shortlisted
1	Energy savings (%)	3.76	1	Yes
2	Normalised energy savings (kWh/m^2 year)	3.46	=12	Yes
3	Electricity consumption saving per year (kWh/year)	3.56	=3	Yes
4	Energy payback period (year)	3.44	=15	-
5	Target green building label	3.14	19	-
6	Payback period (year)	3.47	=10	Yes
7	Return on investment (%)	3.36	16	-
8	Internal rate of return (%)	3.15	18	-
9	Investment cost (USD)	3.75	2	Yes
10	Normalised investment cost (USD/m^2)	3.48	=8	Yes
11	Life cycle cost (USD)	3.48	=8	Yes
12	Increase of building value (%)	3.44	=15	-
13	Ratio of actual to target no. of statutory orders removed (%)	3.46	=12	Yes
14	Ratio of actual to target no. of accidents per year reduced (%)	3.45	13	Yes
15	Target indoor air temperature (°C)	3.50	6	Yes
16	Δ Indoor carbon dioxide levels or harmful substances (ppm)	3.56	=3	Yes
17	Target IAQ class	3.51	5	Yes
18	Target workplane illuminance (lux)	3.47	=10	Yes
19	Target equivalent continuous weighted sound pressure level (dBA)	3.27	17	-

To shortlist the most important KPIs for pragmatic use in building retrofit performance evaluation, 3.45 was taken as the cut-off mean rating. This rating, being the mean between 3.14 and 3.76, represents a moderate-to-high importance level. Thus, a total of 13 KPIs, covering all the four performance aspects, were shortlisted (Table 5).

Table 5. Scenarios of cutoff ratings.

Rating	No. of KPIs Included	Aspects Covered
≥3.00 (Moderate importance)	19	4
≥3.40	15	4
≥3.45 (Moderate-to-high importance)	13	4
≥3.50	6	3
≥4 (High importance)	0	0

5.4. Finalised KPIs

The extraction of the most representative KPIs was based on two criteria: the rank of the KPI and the grouping category. Among the 13 KPIs, KPI-1, KPI-2, and KPI-3 can be used to indicate the energy-saving performance of a retrofit project. Among these three KPIs, KPI-1 was ranked the highest by the respondents, meaning that the practitioners regarded this KPI to be the most important, while KPI-2 and KPI-3 were only ranked the 12th and the 3rd, respectively. Therefore, KPI-1, which can cover the representations of KPI-2 or KPI-3, was taken for use.

KPI-9, KPI-10, and KPI-11 were related to the cost evaluation of a retrofit project. Thus, they can be grouped under one category. As KPI-9 was ranked the highest among the KPIs in this category, the other two KPIs were removed from the list. Additionally, when compared with KPI-9, KPI-11 (life cycle cost) is less feasible in practice because cost elements such as operating and maintenance costs, in the long run, could hardly be accurately determined at the time when a retrofit project is implemented [42].

KPI-16 and KPI-17 were related to IAQ. Despite their similar rankings (KPI-16: rank = 3 and KPI-17: rank = 5), KPI-17 (target IAQ class) covers 12 parameters for IAQ assessment and hence is more representative than a single parameter (Δ Indoor carbon dioxide levels or harmful substances (ppm) covered by KPI-16. The 12 parameters (with 10 chemical parameters) for IAQ assessment are carbon dioxide (CO_2) and other pollutants—namely, carbon monoxide (CO), respirable suspended particulates (PM_{10}), nitrogen dioxide (NO_2), formaldehyde (HCHO), total volatile organic compounds (TVOC), mould, radon, and air-borne bacteria. The certification of IAQ class (Good Class or Excellent Class), administered by the Environmental Protection Department of the Hong Kong government [43], is also authoritative.

For KPI-15, it was only covered in the assessment of IAQ class for building projects completed before 2019. Thus, it was an independent indicator. Participants ranked KPI-16 (Δ indoor carbon dioxide levels or harmful substances (ppm): rank = 3) over KPI-15 (target indoor air temperature (°C): rank: 6). The reason for this may be that occupants can adjust themselves (e.g., putting on or off their clothes) to suit the indoor thermal comfort condition, while they can hardly notice the concentration of the carbon dioxide or harmful substances, not to mention removing such substances. Participants may, therefore, perceive KPI-16 as more important than KPI-15 for building retrofits.

The rest of the original KPIs, i.e., KPI-6, KPI-13, KPI-14, KPI-15, and KPI-18, are independent indicators without overlaps. Thus, they were retained on the KPIs list. The original 13 KPIs and the final eight KPIs are shown in Table 6.

Table 6. Original and final KPIs.

Original (13 KPIs)	Final (8 KPIs)
KPI-1: Energy savings (%)	KPI-1 *: Energy savings (%)
KPI-2: Normalised energy savings (kWh/m^2 year)	
KPI-3: Electricity consumption saving per year (kWh/year)	
KPI-6: Payback period (year)	KPI-2 *: Payback period (year)
KPI-9: Investment cost (USD)	KPI-3 *: Investment cost (USD)
KPI-10: Normalised investment cost (USD/m^2)	
KPI-11: Life cycle cost (USD)	
KPI-13: Ratio of actual to target no. of statutory orders removed (%)	KPI-4 *: Ratio of actual to target no. of statutory orders removed (%)
KPI-14: Ratio of actual to target no. of accidents per year reduced (%)	KPI-5 *: Ratio of actual to target no. of accidents per year reduced (%)
KPI-15: Target indoor air temperature (°C)	KPI-6 *: Target indoor air temperature (°C)
KPI-16: Δ Indoor carbon dioxide levels or harmful substances (ppm)	KPI-7 *: Target IAQ class
KPI-17: Target IAQ class	
KPI-18: Target workplane illuminance (lux)	KPI-8 *: Target workplane illuminance (lux)

Note: * indicates the finalised numbering system.

After the foregoing activities (literature review, focus group, and survey), the shortlisted KPIs, belonging to four different aspects (environmental, economic, health and safety, and users' perspective), were determined. Correspondingly, a hierarchy for the evaluation of building retrofit performance is depicted in Figure 1.

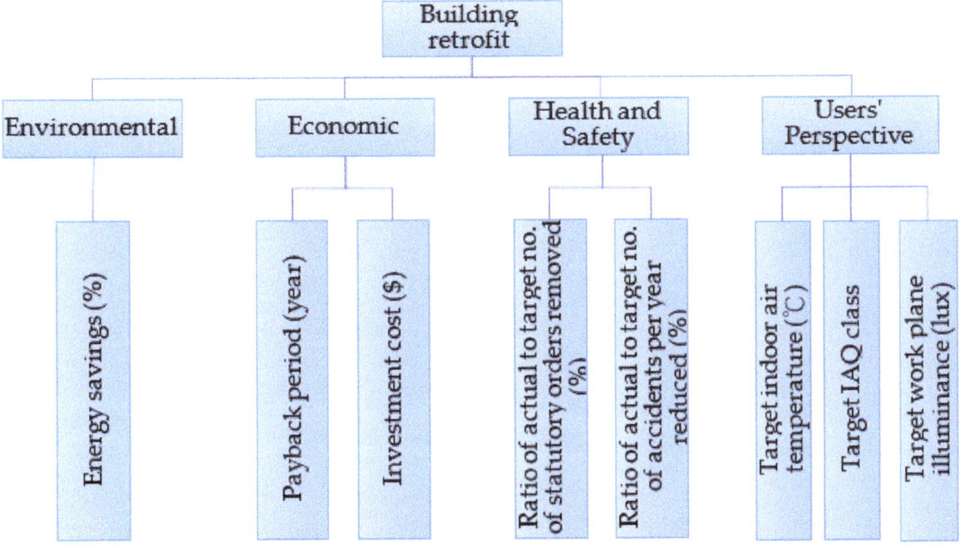

Figure 1. Building retrofit performance hierarchy.

6. Discussion

Part 3 of the survey, containing an open-ended question, asked the participants to provide any comments in relation to the survey topic. From the responses collected, two types of barriers to building retrofits were identified: (1) high evaluation cost and difficulty of obtaining precise cost estimation; (2) different natures of retrofit projects.

6.1. High Evaluation Cost and Difficulty of Obtaining Precise Cost Estimation

The core information to support building retrofit performance assessment is related to safety and proper working of the built assets, health and comfort, space functionality, and energy. Such information is usually gathered by FM managers who arrange maintenance works or technical inspections, or by users who report complaints and fill in satisfaction questionnaires. The main purpose of this is to help improve performance during the operational phase of a building [44]. However, systematic collection of all necessary data to support building retrofit performance evaluation is costly, as one of the survey respondents stated:

'... initial cost (for evaluating the building retrofit performance) can be high ... '

Although data acquisition can be simple using modern and powerful computerised systems [45,46], data overload can be a problem when a sophisticated data mining algorithm is needed to obtain useful information [47]. Thus, whether collecting the data for performance evaluation is worthwhile is a common decision to make for FM managers [45]. If all the applicable KPIs identified from the literature are used to evaluate the performance of building retrofits, considerable effort and resources will be needed to obtain and process the data [37]. Kumar et al. [45] also considered that having many indicators was impractical, and indicators should be simple to allow performance benchmarking [45].

Moreover, it is difficult to accurately predict the energy consumption of commercial buildings. A survey respondent who worked for a private management company stated:

'It (building retrofit) cannot give the precise percentage as the change of the weather might cause the energy consumption to increase significantly. For the energy-saving aspect, we do keep at 2% per year depending on electricity side only.'

As submeters are not commonly installed to monitor the energy consumption of different parts of building services systems [6]. The additional cost of installing submeters, for example for a centralised chiller plant being retrofitted, is usually high given the need to modify the relevant part of the existing system [48]. Therefore, measuring the actual energy saving with respect to the retrofitted portion of the centralised system could be difficult.

6.2. Different Natures of Retrofit Projects

It is common to upgrade the existing equipment at the end of the equipment life or when it comes to failure. Traditional retrofits practice focuses on replacing particular equipment such as chillers and lighting, instead of maximising overall building performance [49]. Additionally, the initial physical condition of the substituting equipment is often emphasised. This is consistent with the following opinion collected from the survey:

'Many retrofit projects initiation was based on the order of equipment or system end of life, no spare part support, change of use/demand, justifiable energy saving whereas functional or environmental enhancement are usually in the lowest priority.'

The following statements from the survey respondents further indicated that the applicability of the KPIs in performance evaluation may vary with the nature and scope of the retrofit projects.

'The ... answers (for KPIs) were generic, in fact, the scores should be depending on the nature of the retrofit project.'

'Some KPIs may not be commonly used or considered by management, and some may not be applicable when designing the retrofit project, while some may be irrelevant to the reason for carrying out retrofit.'

'Special attention should be taken in case of the retrofit project carrying out phase by phase as the newly added and the existing system may be connected and worked together at the same time. Final commissioning of the whole system is necessary at the final stage of the project.'

From the managerial perspective, building performance depends on the resources (e.g., financial, technological, and labour) that are available and the quality of service that should be achieved [50]. In any case, facilities must be assessed with an organisation's goal and mission in place, and the assessment result should inform how well the facilities help the organisation meet its goal and fulfil the mission [51,52].

7. Conclusions

In this study, an industry-wide online survey was conducted to solicit the opinions of FM professionals on the KPIs that are applicable to commercial building retrofit performance evaluation. The results showed that the professionals were generally positive about the importance of the 19 KPIs (with mean ratings being 3 (moderate) or above on a five-point Likert scale). The survey data were analysed through conducting U and H tests, as well as Spearman's rank correlation; the analysed results revealed the variations in the importance of the KPIs perceived by the different groups of professionals.

To enable an effective performance evaluation of building retrofits, the 19 KPIs were further shortlisted. Aside from rating the importance levels of the KPIs, the survey respondents were also invited to provide any other comments about KPIs for building retrofit performance evaluation. Built upon such comments (qualitative) and the KPIs' importance levels (quantitative), a final list of eight KPIs was determined. These KPIs are able to critically reflect the performance of building retrofits in the environmental (energy) aspect and also three other aspects (economic, health and safety, and users' perspective).

Overall, this study contributes to the identification of pragmatic KPIs for commercial building retrofit performance evaluation and serves as a keystone for further development of an analytic evaluation scheme for the assessment of building retrofit performance. Using these KPIs, case studies can be further conducted to examine their applicability in assessing retrofit performance for commercial buildings. To this end, the assessment weighting of each of the KPIs needs to be determined, for example, by an analytic hierarchy process or an analytical network process [53,54], for which the data required can be solicited through interviews with experts working on building retrofits. Furthermore, the methodology of this study can serve as a reference for similar KPI studies in other research domains.

Author Contributions: Conceptualisation, M.Y.H. and J.H.K.L.; methodology, M.Y.H. and J.H.K.L.; validation, M.Y.H., J.H.K.L., H.H. and D.Z.; formal analysis, M.Y.H. and J.H.K.L.; investigation, M.Y.H.; data curation, M.Y.H.; writing—original draft preparation, M.Y.H., J.H.K.L. and H.H.; writing—review and editing, M.Y.H., J.H.K.L., H.H. and D.Z.; visualisation, M.Y.H.; supervision, J.H.K.L.; project administration, M.Y.H. and J.H.K.L. All authors have read and agreed to the published version of the manuscript.

Funding: This research received no external funding.

Institutional Review Board Statement: Not applicable.

Informed Consent Statement: Not applicable.

Acknowledgments: The work reported above is supported by the Department of Building Environment and Energy Engineering of The Hong Kong Polytechnic University.

Conflicts of Interest: The authors declare no conflict of interest.

Appendix A

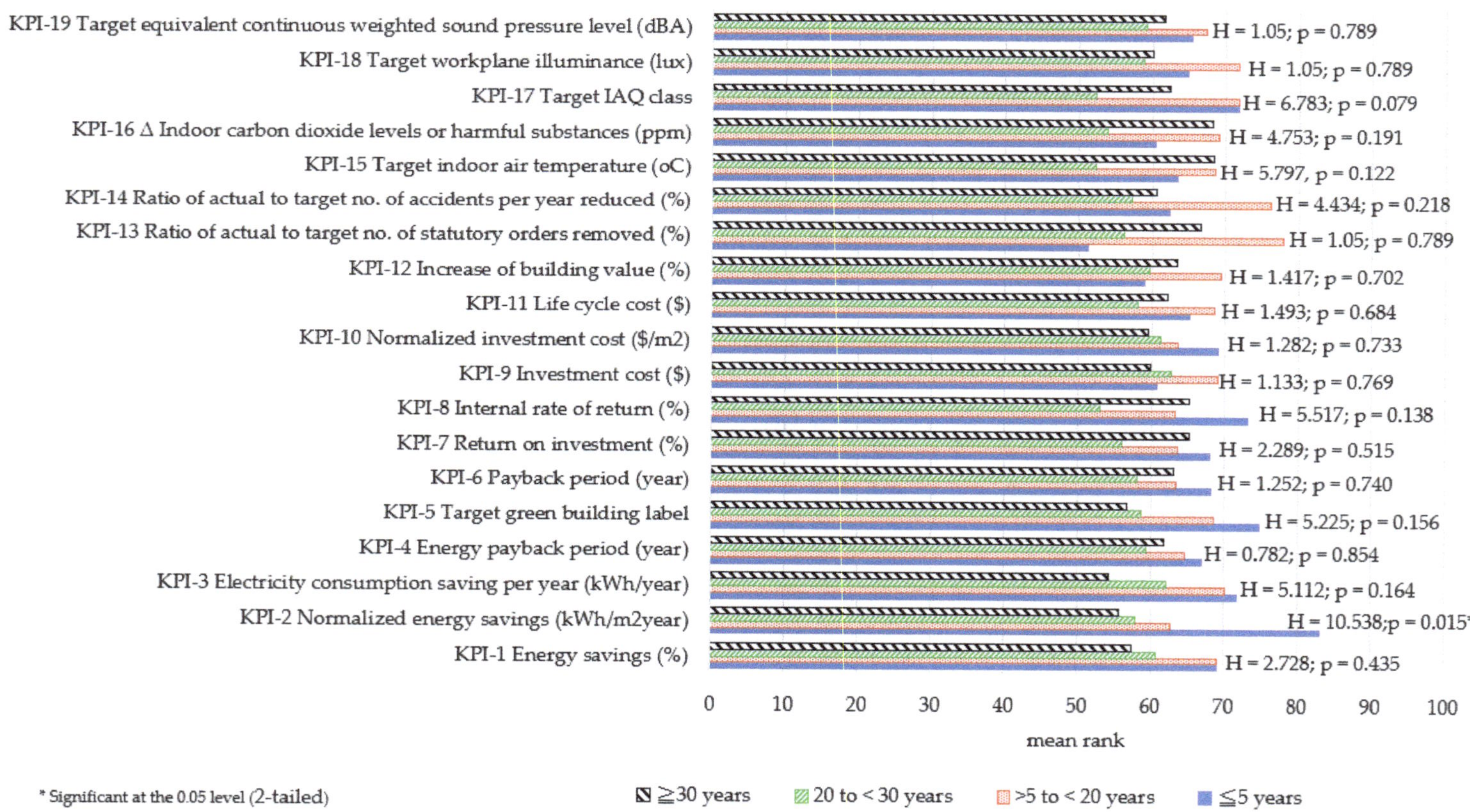

Figure 1. Kruskal–Wallis H-test results of G2 (work experience).

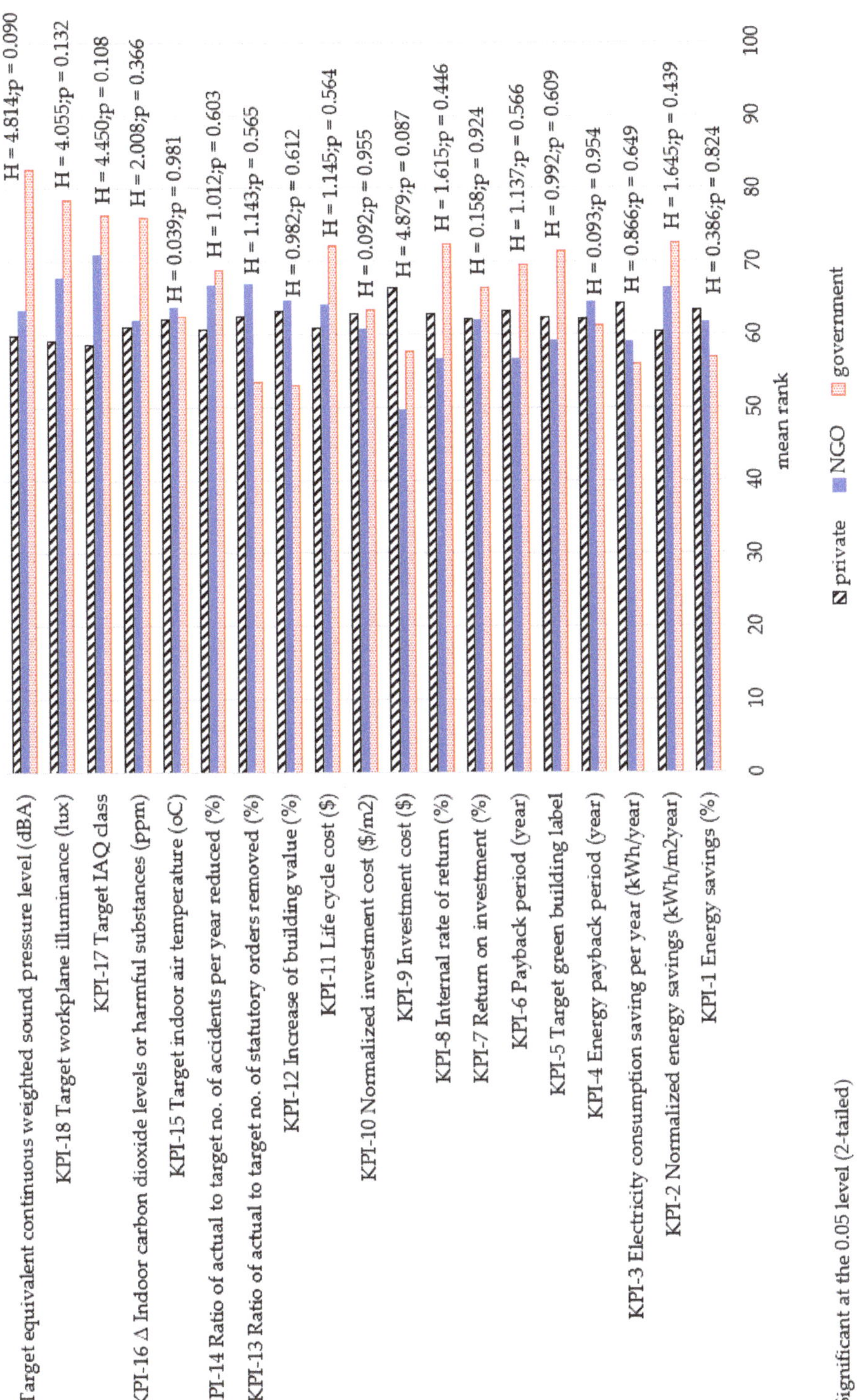

* Significant at the 0.05 level (2-tailed)

Figure 2. Kruskal–Wallis H-test results of G3 (nature of organisation).

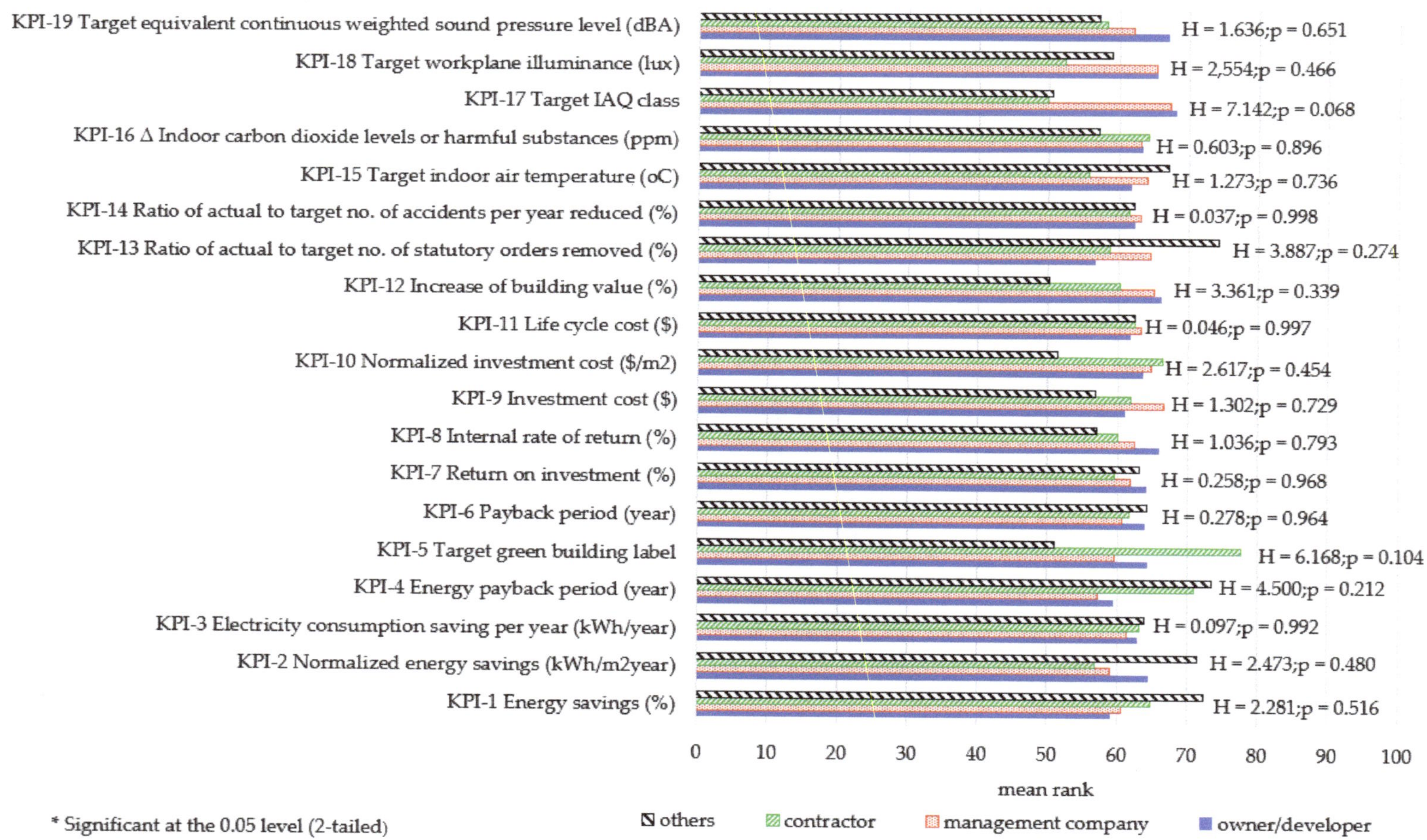

Figure 3. Kruskal–Wallis H-test results of G4 (type of employer).

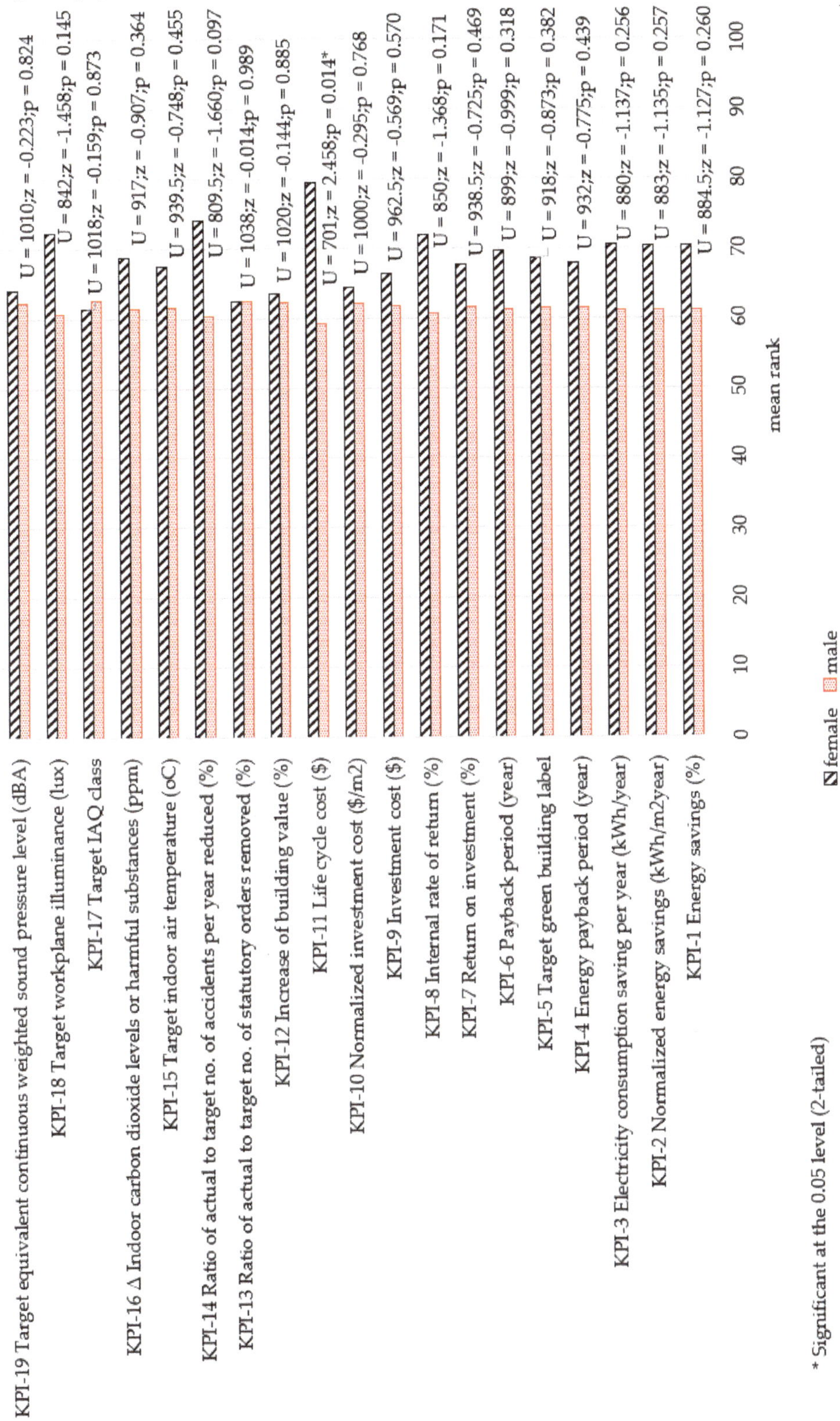

Figure 4. Mann–Whitney U-test results of G1 (gender).

* Significant at the 0.05 level (2-tailed)

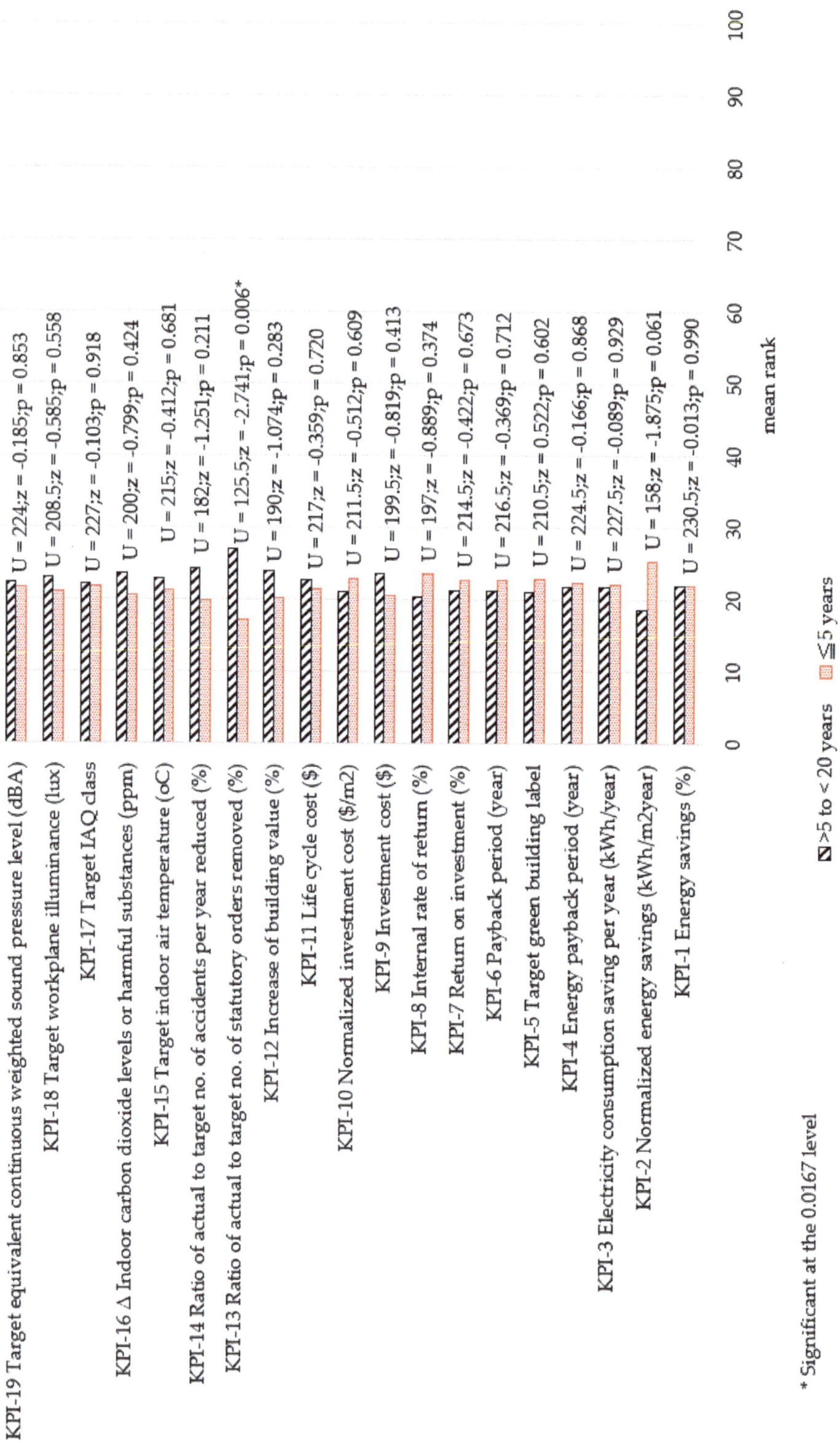

* Significant at the 0.0167 level

Figure 5. Mann–Whitney U-test results of G2 (work experience: G2a vs. G2b).

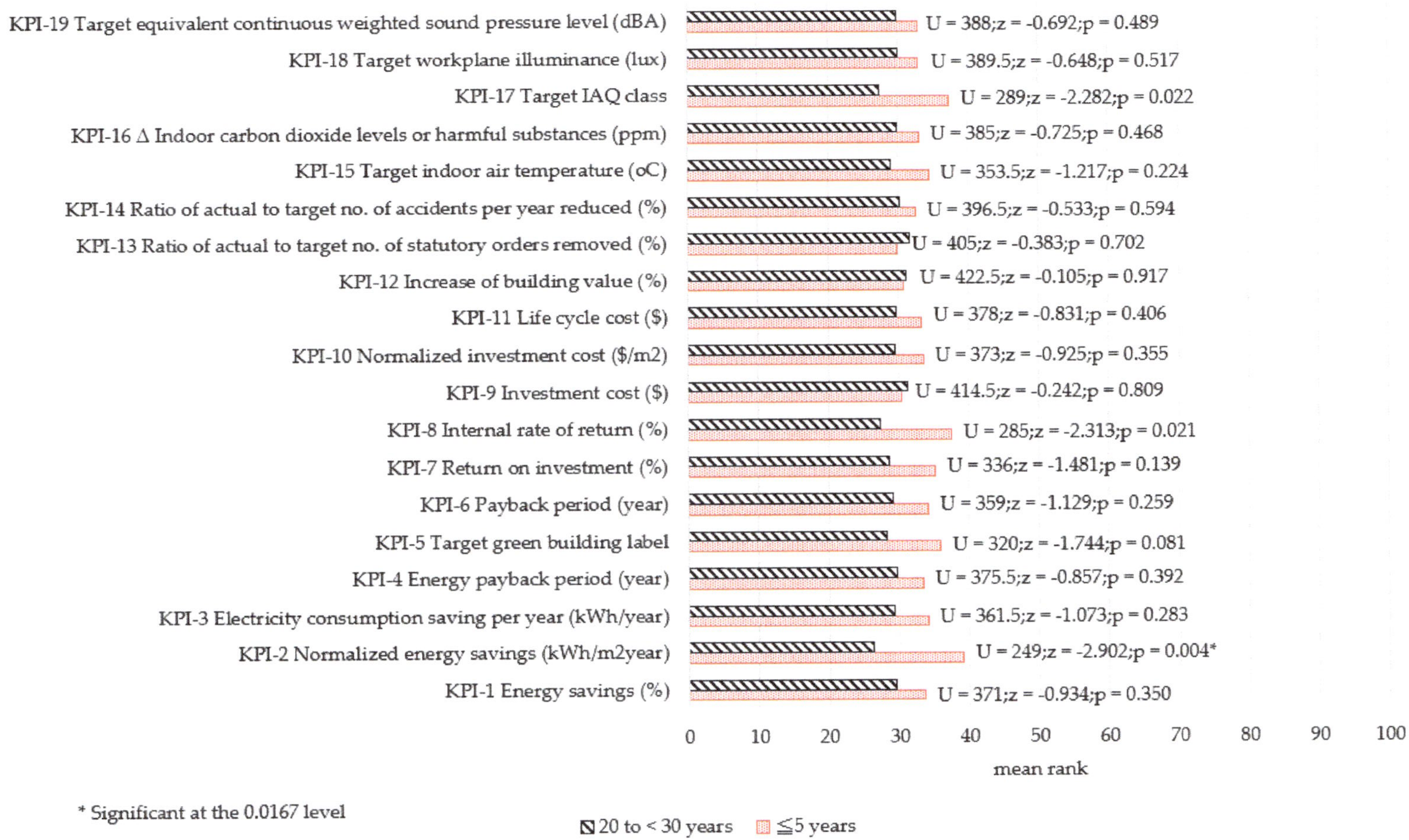

Figure 6. Mann–Whitney U-test results of G2 (work experience: G2a vs. G2c).

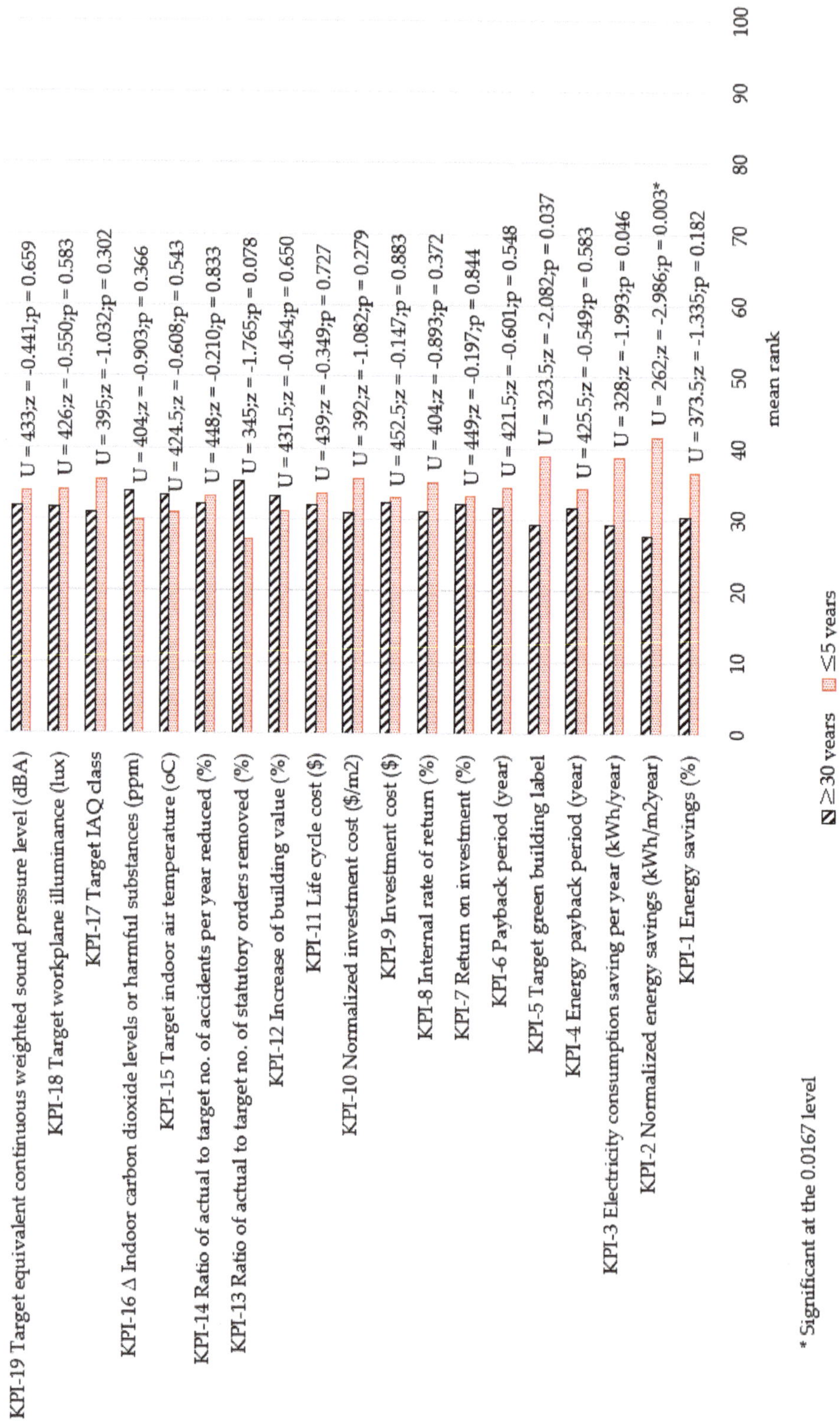

* Significant at the 0.0167 level

Figure 7. Mann–Whitney U-test results of G2 (work experience: G2a vs. G2d).

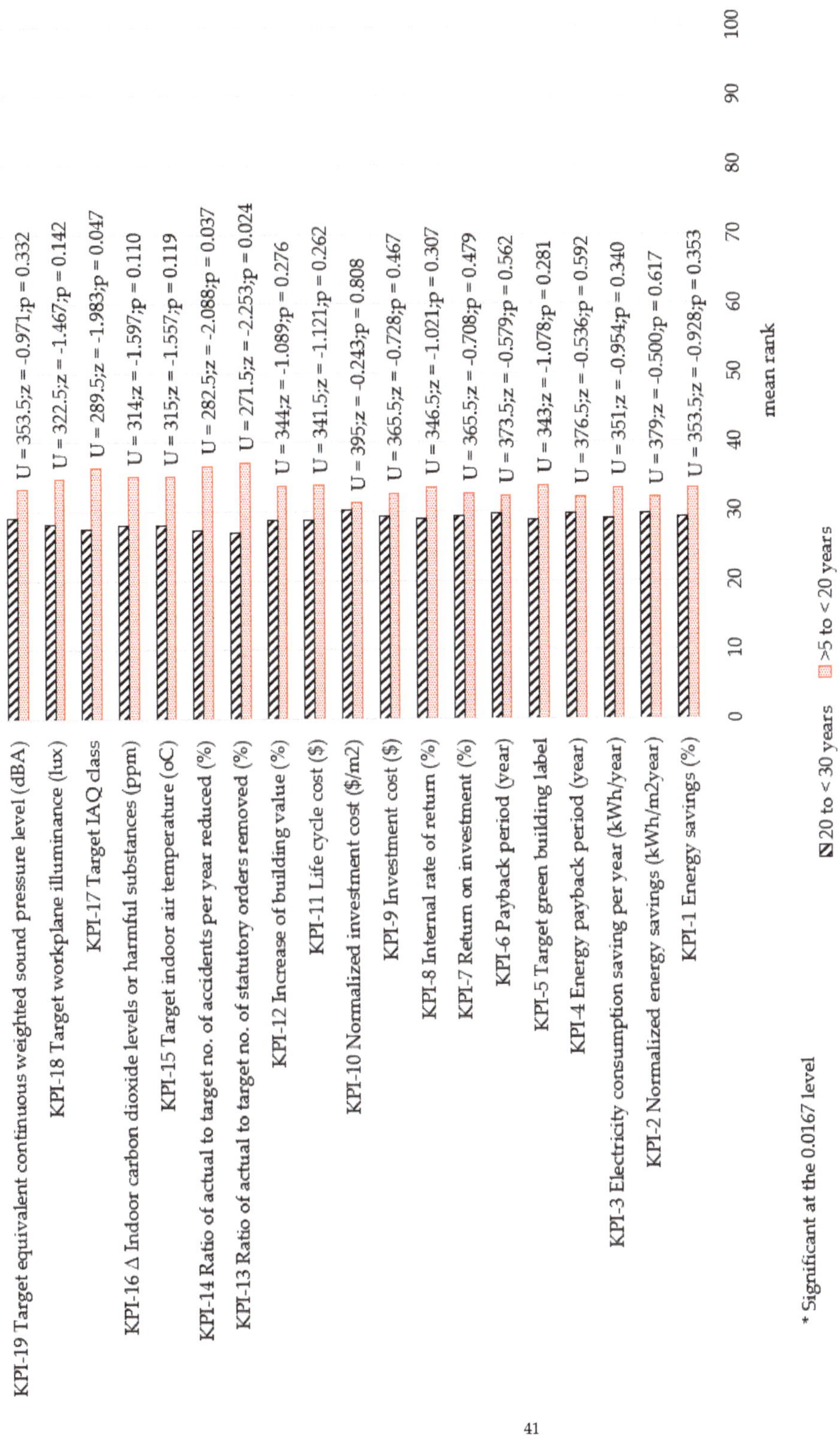

* Significant at the 0.0167 level

Figure 8. Mann–Whitney U-test results of G2 (work experience: G2b vs. G2c).

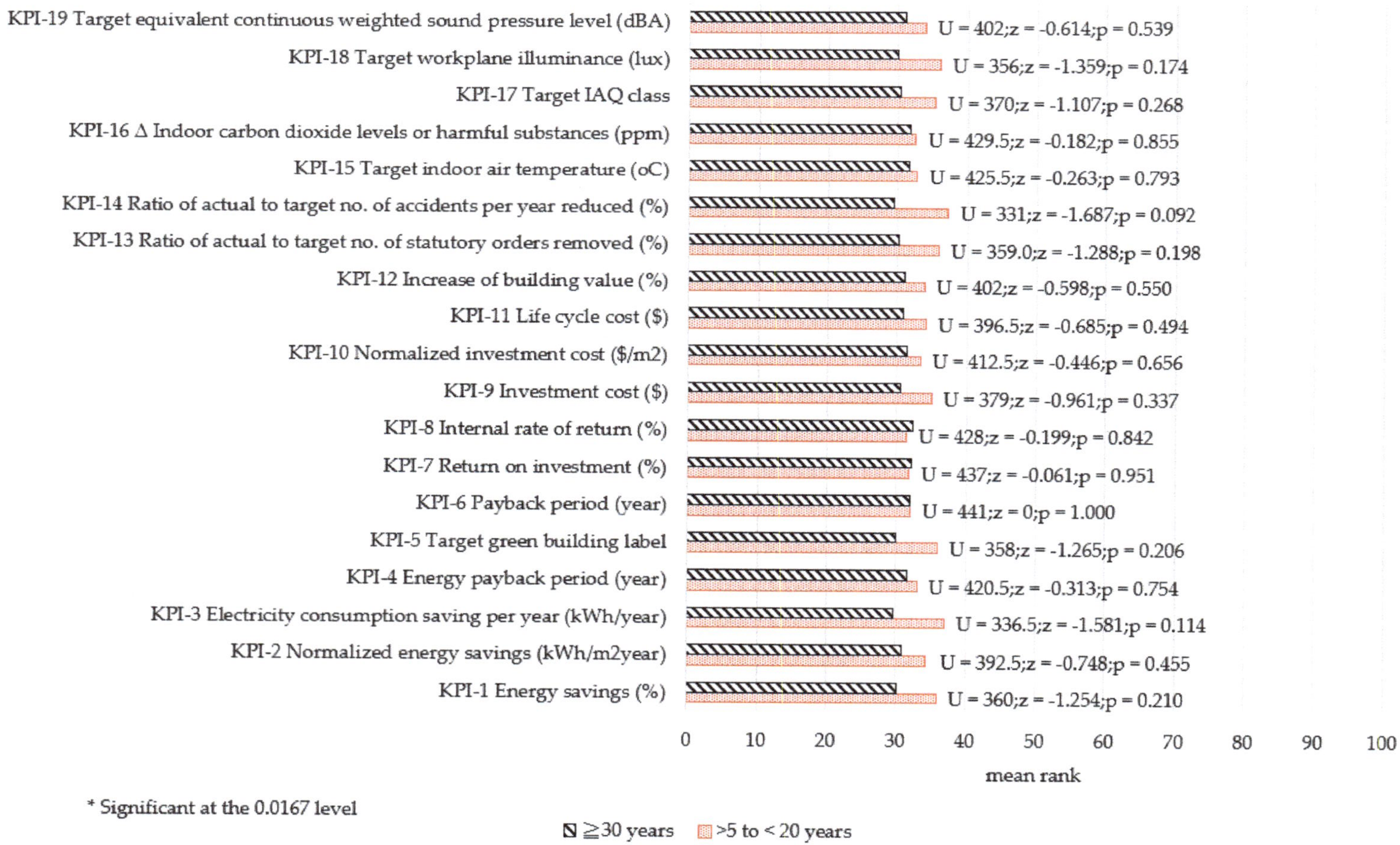

* Significant at the 0.0167 level

Figure 9. Mann–Whitney U-test results of G2 (work experience: G2b vs. G2d).

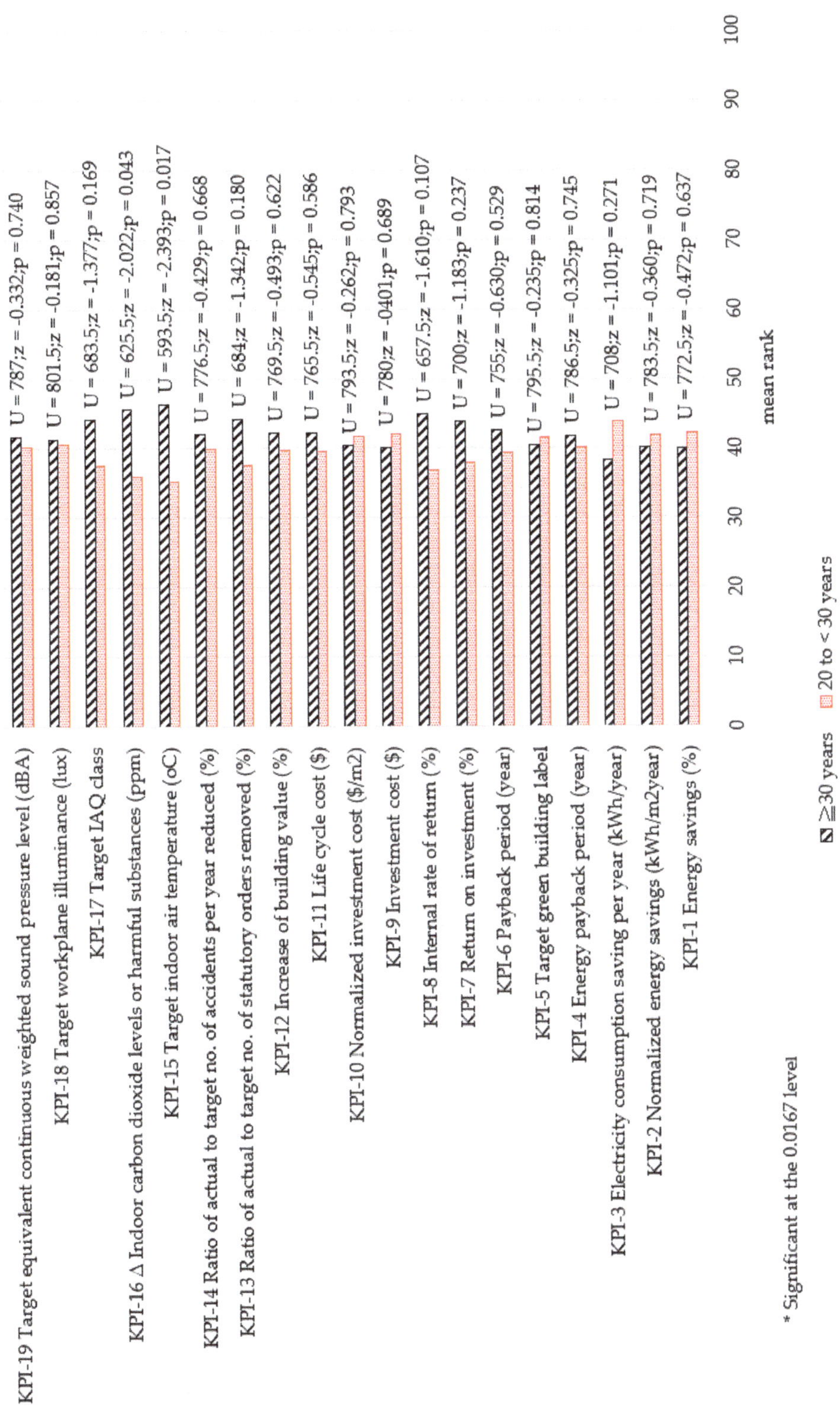

* Significant at the 0.0167 level

Figure 10. Mann–Whitney U-test results of G2 (work experience: G2c vs. G2d).

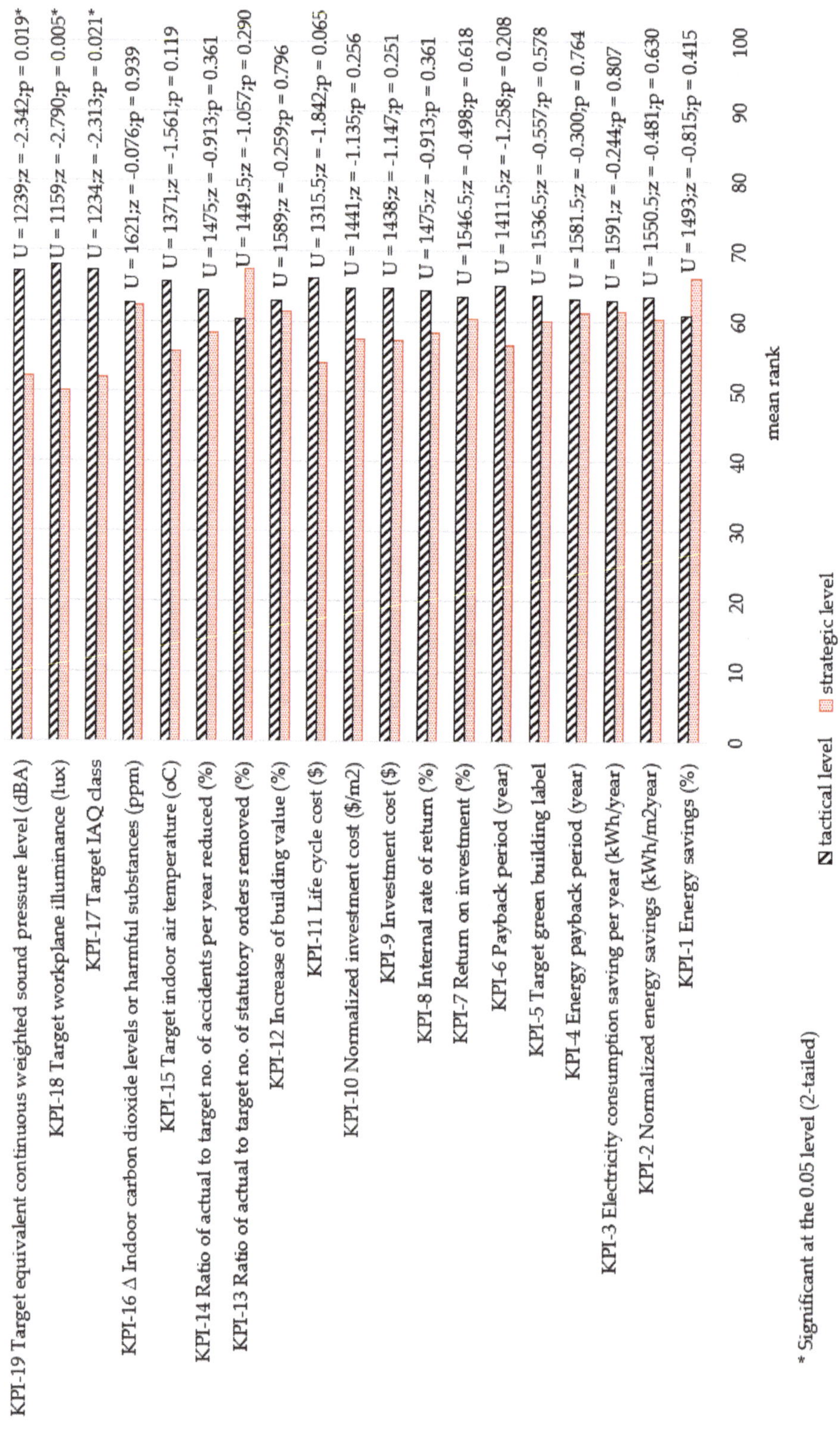

* Significant at the 0.05 level (2-tailed)

Figure 11. Mann–Whitney U-test results of G5 (job level).

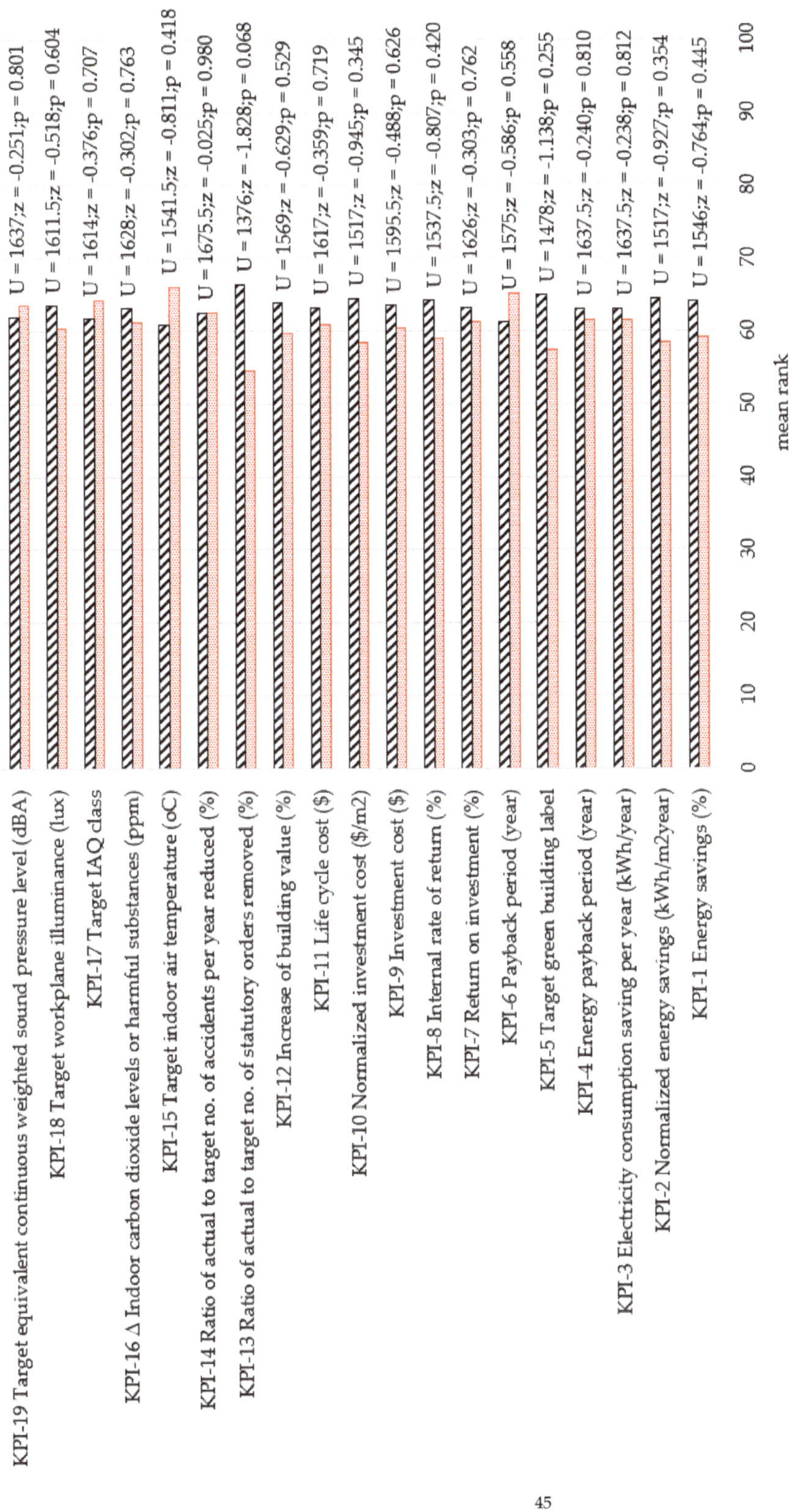

postgraduate degree   sub-degree or bachelor degree

* Significant at the 0.05 level (2-tailed)

Figure 12. Mann–Whitney U-test results of G6 (academic qualification).

A.

Table 1. Mean importance ratings and ranks of KPIs (G1 and G2).

	KPI	G1: Gender				G2: Work Experience							
		a. Male		b. Female		a. ≤5 Years		b. >5 to <20 Years		c. 20 to <30 Years		d. ≥30 Years	
		Mean	Rank	Mean	Rank	Mean	Rank	Mean	Rank	Mean	Rank	Mean	Rank
1	Energy savings (%)	3.70	2	3.95	1	3.95	2	3.90	=1	3.69	2	3.64	=2
2	Normalised energy savings (kWh/m^2 year)	3.40	=12	3.65	10	4.00	1	3.48	=12	3.33	11	3.29	16
3	Electricity consumption saving per year (kWh/year)	3.50	4	3.80	=4	3.86	3	3.76	=4	3.54	3	3.33	15
4	Energy payback period (year)	3.38	14	3.60	=11	3.59	=8	3.48	=12	3.36	=9	3.40	=10
5	Target green building label	3.07	19	3.40	=17	3.59	=8	3.33	18	2.97	18	2.95	19
6	Payback period (year)	3.40	=12	3.70	=7	3.68	=6	3.48	=12	3.31	=12	3.50	5
7	Return on investment (%)	3.30	16	3.55	=13	3.59	=8	3.38	=16	3.18	17	3.40	=10
8	Internal rate of return (%)	3.09	18	3.40	=17	3.45	17	3.19	19	2.87	19	3.24	17
9	Investment cost (USD)	3.72	1	3.85	3	3.73	5	3.90	=1	3.74	1	3.69	1
10	Normalised investment cost (USD/m^2)	3.46	=6	3.60	=11	3.68	=6	3.52	11	3.49	4	3.36	=12
11	Life cycle cost (USD)	3.41	11	3.90	2	3.55	=13	3.62	=8	3.38	=6	3.48	=7
12	Increase of building value (%)	3.42	=9	3.50	15	3.36	19	3.62	=8	3.36	=9	3.48	=7
13	Ratio of actual to target no. of statutory orders removed (%)	3.46	=6	3.45	16	3.49	16	3.47	15	3.44	5	3.49	6
14	Ratio of actual to target no. of accidents per year reduced (%)	3.38	14	3.80	=4	3.55	=13	3.81	3	3.31	=12	3.36	=12
15	Target indoor air temperature (°C)	3.44	8	3.70	9	3.59	=8	3.57	10	3.28	=14	3.62	4
16	Δ Indoor carbon dioxide levels or harmful substances (ppm)	3.52	3	3.75	6	3.55	=13	3.71	6	3.38	=6	3.64	=2
17	Target IAQ class	3.48	5	3.55	=13	3.82	4	3.76	=4	3.28	=14	3.43	9
18	Target workplane illuminance (lux)	3.42	=9	3.70	=7	3.59	=8	3.67	7	3.38	=6	3.36	=12
19	Target equivalent continuous weighted sound pressure level (dBA)	3.27	17	3.35	19	3.41	18	3.38	=16	3.21	16	3.21	18

Table 2. Mean importance ratings and ranks of KPIs (G3 and G4).

	KPI	G3: Nature of Organisation						G4: Type of Employer							
		a. Government		b. NGO		c. Private		a. Owner/ Developer		b. Management Company		c. Contractor		d. Others	
		Mean	Rank	Mean	Rank	Mean	Rank	Mean	Rank	Mean	Rank	Mean	Rank	Mean	Rank
1	Energy savings (%)	3.55	=9	3.78	1	3.78	2	3.70	2	3.70	2	3.83	1	3.95	1
2	Normalised energy savings (kWh/m^2 year)	3.64	=6	3.57	=4	3.41	=12	3.51	=8	3.39	=13	3.33	14	3.63	=3
3	Electricity consumption saving per year (kWh/year)	3.36	=16	3.48	11	3.61	3	3.58	=4	3.52	=6	3.61	=3	3.58	6
4	Energy payback period (year)	3.36	=16	3.57	=4	3.41	=12	3.4	16	3.27	=16	3.61	=3	3.74	2
5	Target green building label	3.45	=12	3.09	18	3.11	19	3.23	=18	3.02	19	3.50	8	2.84	19
6	Payback period (year)	3.64	=6	3.35	=16	3.48	=5	3.53	=6	3.39	=13	3.44	11	3.53	7
7	Return on investment (%)	3.45	=12	3.39	15	3.34	16	3.44	=13	3.34	15	3.22	16	3.37	13
8	Internal rate of return (%)	3.36	=16	3	19	3.17	18	3.23	=18	3.16	18	3.11	=18	3.00	18
9	Investment cost (USD)	3.64	=6	3.48	11	3.83	1	3.72	1	3.82	1	3.78	2	3.63	=3
10	Normalised investment cost (USD/m^2)	3.55	=9	3.48	11	3.48	=5	3.47	=10	3.55	5	3.61	=3	3.26	14
11	Life cycle cost (USD)	3.73	=4	3.48	11	3.46	10	3.44	15	3.5	=8	3.56	=6	3.47	=8
12	Increase of building value (%)	3.27	19	3.52	=7	3.44	11	3.53	=6	3.5	=8	3.39	=12	3.16	17
13	Ratio of actual to target no. of statutory orders removed (%)	3.47	=9	3.48	11	3.48	=5	3.46	12	3.48	=10	3.49	9	3.46	12
14	Ratio of actual to target no. of accidents per year reduced (%)	3.73	=4	3.52	=7	3.40	15	3.44	=13	3.48	=10	3.39	=12	3.47	=8
15	Target indoor air temperature (°C)	3.45	=12	3.61	3	3.48	=5	3.51	=8	3.52	=6	3.28	15	3.63	=3
16	Δ Indoor carbon dioxide levels or harmful substances (ppm)	3.91	1	3.57	=4	3.51	4	3.58	=4	3.57	4	3.56	=6	3.47	=8
17	Target IAQ class	3.82	=2	3.74	2	3.41	=12	3.65	3	3.64	3	3.17	17	3.21	=15
18	Target workplane illuminance (lux)	3.46	=12	3.47	14	3.47	9	3.47	=10	3.48	=10	3.47	10	3.47	=8
19	Target equivalent continuous weighted sound pressure level (dBA)	3.82	=2	3.35	=16	3.19	17	3.37	17	3.27	=16	3.11	=18	3.21	=15

Table 3. Mean importance ratings and ranks of KPIs (G5 and G6).

	KPI	G5: Job Level				G6: Academic Qualification			
		a. Strategic Level		b. Tactical Level		a. Subdegree or Bachelor		b. Postgraduate Degree	
		Mean	Rank	Mean	Rank	Mean	Rank	Mean	Rank
1	Energy savings (%)	3.82	=2	3.73	1	3.68	=1	3.8	1
2	Normalised energy savings (kWh/m^2 year)	3.55	11	3.42	=8	3.35	=14	3.51	6
3	Electricity consumption saving per year (kWh/year)	3.82	=2	3.45	7	3.53	5	3.58	=3
4	Energy payback period (year)	3.61	8	3.36	13	3.38	=14	3.46	=12
5	Target green building label	3.29	18	3.07	19	3.00	19	3.2	18
6	Payback period (year)	3.74	6	3.35	=14	3.53	5	3.44	15
7	Return on investment (%)	3.47	15	3.31	16	3.3	16	3.39	16
8	Internal rate of return (%)	3.26	19	3.1	18	3.08	18	3.19	19
9	Investment cost (USD)	4.11	1	3.59	2	3.68	=1	3.79	2
10	Normalised investment cost (USD/m^2)	3.79	4	3.35	=14	3.35	=14	3.55	5
11	Life cycle cost (USD)	3.53	=12	3.47	=5	3.45	=10	3.5	=7
12	Increase of building value (%)	3.53	=12	3.41	10	3.38	=12	3.48	=9
13	Ratio of actual to target no. of statutory orders removed (%)	3.49	14	3.46	4	3.47	=8	3.46	=12
14	Ratio of actual to target no. of accidents per year reduced (%)	3.58	=9	3.40	=11	3.45	=10	3.45	14
15	Target indoor air temperature (°C)	3.68	7	3.42	=8	3.55	3	3.48	=9
16	Δ Indoor carbon dioxide levels or harmful substances (ppm)	3.58	=9	3.55	3	3.5	7	3.58	=3
17	Target IAQ class	3.76	5	3.40	=11	3.53	5	3.5	=7
18	Target workplane illuminance (lux)	3.46	16	3.47	=5	3.47	=8	3.47	11
19	Target equivalent continuous weighted sound pressure level (dBA)	3.37	17	3.23	17	3.25	17	3.29	17

References

1. IEA. World Energy Statistics and Balances (Database). Available online: www.iea.org/statistics (accessed on 31 July 2021).
2. Tan, Y.; Liu, G.; Zhang, Y.; Shuai, C.; Shen, G.Q. Green retrofit of aged residential buildings in Hong Kong: A preliminary study. *Build. Environ.* **2018**, *143*, 89–98. [CrossRef]
3. Zheng, L.; Lai, J. Environmental and Economic Evaluations of Building Energy Retrofits: Case Study of a Commercial Building. *Build. Environ.* **2018**, *145*, 14–23. [CrossRef]
4. Sing, M.C.P.; Chan, V.W.C.; Lai, J.H.K.; Matthews, J. Energy-efficient retrofitting of multi-storey residential buildings. *Facilities* **2021**, *39*, 722–736. [CrossRef]
5. Ho, A.; Lai, J. Building retrofit: Review on modelling studies, real applications and barriers. In Proceedings of the 9th Great Pearl River Delta and 2nd Guangdong, Hong Kong and Macao Dawan District Building Operation and Maintenance Conference, Guangzhou, China, 15 December 2018; pp. 54–61.
6. Ho, A.M.Y.; Lai, J.H.K.; Chiu, B.W.Y. Key performance indicators for holistic evaluation of building retrofits: Systematic literature review and focus group study. *J. Build. Eng.* **2021**, *43*, 102926. [CrossRef]
7. Shaikh, P.H.; Shaikh, F.; Sahito, A.A.; Uqaili, M.A.; Umrani, Z. Chapter 9-An Overview of the Challenges for Cost-Effective and Energy-Efficient Retrofits of the Existing Building Stock. In *Cost-Effective Energy-Efficient Building Retrofitting*, 1st ed.; Fernando, P., Claes-Göran, G., Bjørn, P.J., Giuseppe, P.V., Nicola, B., Jarek, K., Eds.; Woodhead Publishing: Cambridge, UK, 2017; pp. 257–278.
8. Jagarajan, R.; Abdullah, M.; Asmoni, M.N.; Mohammed, A.H.; Jaafar, M.N.; Lee, Y.M.J.; Baba, M. Green retrofitting–A review of current status, implementations and challenges. *Renew. Sustain. Energy Rev.* **2017**, *67*, 1360–1368. [CrossRef]
9. Aghamolaei, R. Evaluation of Supply and Demand in Building Energy Performance: Application of Retrofit Scenarios in Residential Building. *Energy Eng.* **2019**, *116*, 60–79. [CrossRef]
10. Ferrari, S.; Beccali, M. Energy-environmental and cost assessment of a set of strategies for retrofitting a public building toward nearly zero-energy building target. *Sustain. Cities Soc.* **2017**, *32*, 226–234. [CrossRef]
11. Wilkinson, S. Analysing sustainable retrofit potential in premium office buildings. *Struct. Surv.* **2012**, *30*, 398–410. [CrossRef]
12. Gagliano, A.; Detommaso, M.; Nocera, F.; Patania, F.; Aneli, S. The Retrofit of Existing Buildings Through the Exploitation of the Green Roofs–A Simulation Study. *Energy Procedia* **2014**, *62*, 52–61. [CrossRef]
13. Tokede, O.; Ahiaga-Dagbui, D. Evaluating the whole-life cost implication of revocability and disruption in office retrofit building projects. In Proceedings of the 32nd ARCOM Conference, Association of Researchers in Construction Management, Manchester, UK, 5–7 September 2016; pp. 321–330.
14. Jafari, A.; Valentin, V. An optimization framework for building energy retrofits decision-making. *Build. Environ.* **2017**, *115*, 118–129. [CrossRef]
15. Ma, Z.; Cooper, P.; Daly, D.; Ledo, L. Existing building retrofits: Methodology and state-of-the-art. *Energy Build.* **2012**, *55*, 889–902. [CrossRef]
16. Terés-Zubiaga, J.; Martin, K.; Erkoreka, A.; Aparicio, X.; del Portillo, L.A. Chapter 18-Cost-Effective Energy Retrofitting of Buildings in Spain: An Office Building of the University of the Basque Country. In *Cost-Effective Energy Efficient Building Retrofitting*, 1st ed.; Fernando, P., Claes-Göran, G., Bjørn, P.J., Giuseppe, P.V., Nicola, B., Jarek, K., Eds.; Woodhead Publishing: Cambridge, UK, 2017; pp. 515–551.
17. Remer, D.S.; Nieto, A.P. A compendium and comparison of 25 project evaluation techniques. Part 1. Net present value and rate of return methods. *Int. J. Prod. Econ.* **1995**, *42*, 79–96. [CrossRef]
18. Verbeeck, G.; Hens, H. Energy savings in retrofitted dwellings: Economically viable. *Energy Build.* **2005**, *37*, 747–754. [CrossRef]
19. Nikolaidis, Y.; Pilavachi, P.A.; Chletsis, A. Economic evaluation of energy saving measures in a common type of Greek building. *Appl. Energy* **2009**, *86*, 2550–2559. [CrossRef]
20. Peterson, S.; Svendsen, S. Method for component-based economical optimisation for use in design of new low–energy buildings. *Renew. Energy* **2012**, *38*, 173–180. [CrossRef]
21. Ornetzeder, M.; Wicher, M.; Suschek-Berger, J. User satisfaction and well-being in energy efficient office buildings: Evidence from cutting-edge projects in Austria. *Energy Build.* **2016**, *118*, 18–26. [CrossRef]
22. Beccali, M.; Bonomolo, M.; Lo Brano, V.; Ciulla, G.; Di Dio, V.; Massaro, F.; Favuzza, S. Energy saving and user satisfaction for a new advanced public lighting system. *Energy Convers.* **2019**, *195*, 943–957. [CrossRef]
23. Kylili, A.; Paris, A. Fokaides, and Petra Amparo Lopez Jimenez. Key Performance Indicators (KPIs) approach in buildings renovation for the sustainability of the built environment: A review. *Renew. Sustain. Energy Rev.* **2016**, *56*, 906–915. [CrossRef]
24. Gourlis, G.; Kovacic, I. A study on building performance analysis for energy retrofit of existing industrial facilities. *Appl. Energy* **2016**, *184*, 1389–1399. [CrossRef]
25. Walter, T.; Sohn, M.D. A regression-based approach to estimating retrofit savings using the Building Performance Database. *Appl. Energy* **2016**, *179*, 996–1005. [CrossRef]
26. Cuerda, E.; Guerra-Santin, O.; Sendra, J.J.; Neila, F.J. Understanding the performance gap in energy retrofitting: Measured input data for adjusting building simulation models. *Energy Build.* **2020**, *209*, 109688. [CrossRef]
27. Pohoryles, D.A.; Maduta, C.; Bournas, D.A.; Kouris, L.A. Energy performance of existing residential buildings in Europe: A novel approach combining energy with seismic retrofitting. *Energy Build.* **2020**, *223*, 110024. [CrossRef]

28. Seyedzadeh, S.; Rahimian, F.P.; Oliver, S.; Rodriguez, S.; Glesk, I. Machine learning modelling for predicting non-domestic buildings energy performance: A model to support deep energy retrofit decision-making. *Appl. Energy* **2020**, *279*, 115908. [CrossRef]
29. Al Dakheel, J.; Del Pero, C.; Aste, N.; Leonforte, F. Smart buildings features and key performance indicators: A review. *Sustain. Cities Soc.* **2020**, *61*, 102328. [CrossRef]
30. Collins, A.J.; Hester, P.; Ezell, B.; Horst, J. An improvement selection methodology for key performance indicators. *Environ. Syst. Decis.* **2016**, *36*, 196–208. [CrossRef]
31. Xu, P.; Chan, E.H.W.; Queena, Q.K. Key performance indicators (KPI) for the sustainability of building energy efficiency retrofit (BEER) in hotel buildings in China. *Facilities* **2012**, *30*, 432–448.
32. Lai, J.H.; Hou, H.C.; Chiu, B.W.; Edwards, D.; Yuen, P.L.; Sing, M.; Wong, P. Importance of hospital facilities management performance indicators: Building practitioners' perspectives. *J. Build. Eng.* **2022**, *45*, 103428. [CrossRef]
33. Luther, M.B.; Rajagopalan, P. Defining and developing an energy retrofitting approach. *J. Green Build.* **2014**, *9*, 151–162. [CrossRef]
34. Escrivá-Escrivá, G. Basic actions to improve energy efficiency in commercial buildings in operation. *Energy Build.* **2011**, *43*, 3106–3111. [CrossRef]
35. Mawed, M.; Tilani, V.; Hamani, K. The role of facilities management in green retrofit of existing buildings in the United Arab Emirates. *J. Facil. Manag.* **2020**, *18*, 36–52. [CrossRef]
36. Lai, J.H.K.; Man, C.S. Performance indicators for facilities operation and maintenance (Part 1): Systematic classification and mapping. *Facilities* **2018**, *36*, 476–494. [CrossRef]
37. Lai, J.H.K.; Man, C.S. Performance indicators for facilities operation and maintenance (Part 2): Shortlisting through a focus group study. *Facilities* **2018**, *36*, 495–509. [CrossRef]
38. Lai, J.H.K.; Yuen, P.L. Identification, classification and shortlisting of performance indicators for hospital facilities management. *Facilities* **2021**, *39*, 4–18. [CrossRef]
39. Corder, G.W.; Foreman, D.I. *Nonparametric Statistics: A Step-by-Step Approach*, 2nd ed.; Wiley: Hoboken, NJ, USA, 2014.
40. Rodríguez, P.; Montequín, V.R.; Morán, H.; de Arriba, L. Gender influence in project management: Analysis of a case study based on master students. *Procedia Comput. Sci.* **2017**, *121*, 461–468. [CrossRef]
41. Miller, A.; Higgins, C. Deep Energy Savings in Existing Buildings. *ASHRAE Trans.* **2015**, *121*, 380–394.
42. Lai, J.H.K. Operation and maintenance budgeting for commercial buildings in Hong Kong. *Constr. Manag. Econ.* **2010**, *28*, 415–427. [CrossRef]
43. Environmental Campaign Committee and Environmental Protection Department. Hong Kong Green Organisation Certification-Guidebook for IAQwi$e Certificate. Available online: https://www.hkgoc.gov.hk/uploads/2021%20IAQ-Guidebook_Eng_Clean.pdf (accessed on 16 August 2021).
44. Bortolini, R.; Forcada, N. Facility managers' perceptions on building performance assessment. *Front. Eng. Manag.* **2018**, *5*, 324–333.
45. Kumar, U.; Galar, D.; Parida, A.; Stenström, C.; Berges, L. Maintenance performance metrics: A state-of-the-art review. *J. Qual. Maint. Eng.* **2013**, *19*, 233–277. [CrossRef]
46. Lai, J.H.K. Maintenance performance: Examination of the computer-aided maintenance data of a large commercial building. *J. Perform. Constr. Facil.* **2015**, *29*, 4014118. [CrossRef]
47. Charnes, A.; Clark, C.T.; Cooper, W.W.; Golany, B. A developmental study of data envelopment analysis in measuring the efficiency of maintenance units in the US air forces. *Ann. Oper. Res.* **1984**, *2*, 95–112. [CrossRef]
48. Lai, J.; Yik, F.; Jones, P. Expenditure on operation and maintenance service and rental income of commercial buildings. *Facilities* **2008**, *26*, 242–265. [CrossRef]
49. Amann, J.T.; Mendelsohn, E. *Comprehensive Commercial Retrofit Programs: A Review of Activity and Opportunities*; American Council for an Energy-Efficient Economy: Washington, DC, USA, 2005.
50. Preiser, W.; Nasar, J. Assessing Building Performance: Its Evolution from Post-Occupancy Evaluation. *Int. J. Archit. Res.* **2008**, *2*, 84–99.
51. National Research Council; John, H.C.; Federal Facilities Council Ad Hoc; Committee on Performance Indicators for Federal Real Property Asset Management; Jocelyn, S.D.; Division on Engineering Physical Sciences; Federal Facilities Council. *Key Performance Indicators for Federal Facilities Portfolios*; National Academies Press Publications Sales Office: Washington, DC, USA, 2005.
52. Cripps, S. Calgary K-12 district benefits from the facilities management evaluation program. *Facil. Manag.* **1998**, *14*, 27,28,31,33.
53. Saaty, T.L. *The Analytic Hierarchy Process*; McGraw-Hill: New York, NY, USA, 1980.
54. Saaty, T.L. *Theory and Applications of the Analytic Network Process: Decision Making with Benefits, Opportunities, Costs, and Risks*; RWS Publications: Pittsburgh, PA, USA, 2005.

energies

Article

Case Study and Feasibility Analysis of Multi-Objective Life Cycle Energy System Optimization in a Nordic Campus Building

Vilppu Eloranta [1,*], Aki Grönman [2] and Aleksandra Woszczek [2]

[1] Faculty of Technology, LAB University of Applied Sciences, Mukkulankatu 19, 15210 Lahti, Finland
[2] School of Energy Systems, LUT University, Yliopistonkatu 34, 53850 Lappeenranta, Finland; aki.gronman@lut.fi (A.G.); aleksandra.woszczek@lut.fi (A.W.)
* Correspondence: vilppu.eloranta@lab.fi

Abstract: Due to the high energy consumption of buildings, there is a demand for both economically and environmentally effective designs for building energy system retrofits. While multi-objective optimization can be used to solve complicated problems, its use is not yet widespread in the industry. This study first aims to develop an efficient and applicable multi-objective building energy system optimization method, used to dimension energy production and storage retrofit components in a case campus building in Lahti, Finland. Energy consumption data of the building are obtained with a dynamic energy model. The optimization model includes economic and environmental objectives, and the approach is found to function satisfactorily. Second, this study aims to assess the feasibility and issues of multi-objective single-building energy system optimization via the analysis of the case optimization results. The results suggest that economically beneficial local energy production and storage retrofits could not always lead to life cycle CO_2-eq emission reductions. The recognized causes are high life cycle emissions from the retrofit components and low Nordic grid energy emissions. The performed sensitivity and feasibility analyses show that correctness and methodological comparability of the used emission factors and future assumptions are crucial for reliable optimization results.

Keywords: building energy system optimization; renewable energy retrofit; life cycle emission

Citation: Eloranta, V.; Grönman, A.; Woszczek, A. Case Study and Feasibility Analysis of Multi-Objective Life Cycle Energy System Optimization in a Nordic Campus Building. *Energies* **2021**, *14*, 7742. https://doi.org/10.3390/en14227742

Academic Editors: Cynthia Hou and Gianpiero Colangelo

Received: 5 October 2021
Accepted: 15 November 2021
Published: 18 November 2021

1. Introduction

Buildings and their construction consume over one-third of the total global energy consumption and cause almost 40% of the global carbon dioxide (CO_2) emissions [1], and 36% of the CO_2 emissions in the European Union (EU). The EU Energy Performance of Buildings Directive (EPBD) already requires new buildings to be nearly zero-energy buildings (NZEB); however, the EU building stock is also rather old and slowly renovated. Therefore, retrofits and modernizations to existing properties offer large energy performance and sustainability potential [2]. Thus, there is a need for reliable methods to identify the best-performing retrofit and modernization targets.

In a typical Finnish design, on-site renewable energy production, especially photovoltaic (PV) generation, is dimensioned to maximize self-utilization of the produced energy in the building, and therefore, to minimize the excess production, usually exported to the grid [3]. To increase peak renewable energy production, exported energy needs to be increased, demand flexibility needs to be implemented, or storage technologies need to be utilized. If energy storage is added to the system, dimensioning becomes complicated because of component interdependency. The system state is defined not only by current energy flows but also by past energy flows that manifest in storage state-of-charge. Therefore, energy systems containing storage components cannot be easily dimensioned with traditional methods, and the design must often be assisted with simulation.

Dynamic building energy simulation software, such as IDA ICE [4], EnergyPlus [5], and TRNSYS [6], couple the indoor zone conditions and energy system state at each timestep. Dynamic models can include the entire building with its HVAC systems and on-site energy sources, including solar and geothermal. The software can be used for various purposes, such as dimensioning the energy system components, assessing indoor climate conditions, or verifying certifications. However, once the problem becomes truly dynamic and interdependent with storage systems, fully dynamic approaches are computationally intensive. One approach is to draft multiple scenarios with different component sizes, and run the full dynamic simulation for each case [7]. The optimal case may be chosen based on the results. However, this brute force approach requires a lot of manual work and computational resources and might, therefore, not be feasible in some applications.

This problem can be solved through separating the optimization from the dynamic model. Several multi-stage energy system optimization approaches for buildings have been recently studied, utilizing dynamic energy modeling tools such as EnergyPlus [8–11] or IDA ICE [12] to obtain the demand data. Optimization is then performed with separate tools, such as MATLAB, either with or without linking it to the dynamic simulation software. Studied and tested algorithms include linear programming (LP) [13–15], mixed-integer linear programming (MILP) [8,16,17], genetic algorithm (GA) [9,10,12,18], and particle swarm optimization (PSO) [11,19].

Energy system optimization approaches have not yet become widespread in practical system design processes. Challenges of utilizing energy optimization in these workflows have been analyzed via specialist interviews [20] and literature reviews [21]. The interview study [20] found various issues, such as variation of methods, high required expertise, validity concerns, required computation and time resources, lacking or poor-quality input data, and lack of production-ready tools. Nguyen et al. [21] state problems such as missing interfaces between simulation and optimization software, difficulty of adequate trade-offs between accuracy and simplicity, and lack of policies enforcing the use of optimization. Moreover, integration of life cycle emission considerations to building design processes is problematic due to inconsistent values and methods. Identified issues include non-comparable reporting standards, operational vs. embodied emissions, and region-specific and inadequate calculation tools [22].

Due to the old EU building stock and the need to improve its energy performance, energy retrofits and their optimization has been studied broadly. Ibañez Iralde et al. [23] reviewed the current energy retrofit measures and funding for residential buildings in Spain, and specified retrofit possibilities. Galatioto et al. [24] reviewed the feasibilities of energy retrofits in historical Italian buildings, considering insulation, energy storage, and intelligent control actions, among other possibilities. Gagliano et al. [25] compared the configuration scenarios of PV panels, battery capacities and energy consumption profiles in residential applications to minimize energy exchange with the grids. He et al. [26] presented a comprehensive retrofit analysis considering wall insulation, piping system, occupancy sensors, and other means. Nocera et al. [27] performed a retrofit analysis for a hotel in Italy, studying broad actions to achieve nearly zero-energy building (nZEB) status. Pirmohamadi et al. [28] demonstrated an optimization approach for an office solar thermal system in Iran to optimize energy consumption, and economic and environmental costs. These studies, however, take place outside the Nordic conditions characterized by cold climate and relatively low electricity and heat energy prices. Moreover, the introduced studies do not consider life cycle emissions from the retrofit equipment manufacturing.

From these premises, this study first aims to develop a computationally efficient and applicable multi-objective building energy system optimization method for energy system component dimensioning, utilizing balanced economic and environmental optimization objectives. The approach is based on linear optimization that can be applied in early design phases for either new construction or retrofit and requires either simulated or measured energy demand data of the building as an input. It extends previous research by developing a modular and user-friendly template, utilizing built-in components from the open-source energy system modeling framework Calliope [29]. The approach is used to dimension energy system component retrofits for a campus building in Lahti, Finland,

and environmental life cycle optimization is enabled by obtaining component life cycle emission data from literature. Second, this study aims to assess the feasibility of economic-environmental multi-objective single-building energy system optimization by analyzing result reliability and comparability. The process is enabled by sensitivity, feasibility, and qualitative analyses of the case optimization results, from which potential actions and best practices are derived and discussed.

The article is organized as follows: first, the applied modeling approaches are presented, including the model validation and used parameters. Second, the optimization results are presented along with sensitivity analyses and further insights. Last, the results are discussed along with future research needs and conclusions.

2. Methodology

The case building in Lahti, Finland is a repurposed furniture factory, currently accommodating production, office, storage, and campus spaces. Floor plans of the building are available in Appendix A. The building was originally built in the 1950s and partly modernized in the late 2010s, when 111 kW photovoltaic (PV) solar panels and a 1 MW ground source heat pump (GSHP) system with ground boreholes were installed. The building is connected to the local district heat (DH) network.

As the building consists of several parts and is partially renovated, it contains structural parts from different eras, differing in heat conductivity. Original windows have double glazing, i.e., inner and outer frames with single panes, whereas modernized windows have triple glazing, i.e., inner frames with two pane insulating glass elements and outer frames with single panes. Large glazing areas have solar control glass.

Due to the northern location of Finland, the achievable PV production profile is distinct with high insolation in the long summer days, and low in the short winter days [30]. Hence, utilizing excess PV electricity production in the summer to drive heat pumps and storing the generated heat seasonally could be one solution to balance this characteristic. Combining PV production and heat pump with battery [31] or thermal storage [32] have been studied; however, the integration of these systems in Nordic conditions still needs further research. The feasibility of this concept is investigated in the results.

As this study focuses exclusively on the building energy system, other means of energy performance improvement, such as renovation, insulation improvement, occupant coaching, or schedule optimization, are not considered.

2.1. Dynamic Building Energy Model and Validation

Energy demand data for the building were obtained with dynamic building simulation software IDA Indoor Climate and Energy (IDA ICE) [4]. The building geometry was derived from floor plans and by site inspection. Thermodynamic properties, such as thermal conductivities, were sourced from structural design documentation and by inspection, and they are presented in Table 1. The model floor area is 79,184 m^2 and the volume 374,643 m^3.

Table 1. Areas and heat conductivities of the building structures in the dynamic energy model. Heat conductivity is presented as the average value with the range of various structures in parentheses.

Structure	Area [m^2]	Heat Conductivity [W/(m^2K)]
External wall	16,250	0.42 (0.17–0.66)
Roof	25,387	0.31 (0.09–0.40)
Slab, external floor	25,524	0.56 (0.11–0.59)
Glazing to north	2125	2.02 (1.00–2.90)
Glazing to east	900	1.38 (1.00–2.90)
Glazing to south	1743	2.21 (1.00–2.90)
Glazing to west	2459	1.17 (1.00–2.90)

The default energy plant model of IDA ICE was used, containing simple heating and cooling heat exchangers with unlimited capacities. Heating distribution fluid temperature was set to adjust linearly based on the ambient temperature, from 70 °C at −30 °C ambient to 20 °C at +20 °C ambient, and the zones were heated with water radiators with 21 °C setpoints. Indoor cooling was performed mainly via chilled supply air with 24 °C zone setpoints. The standard IDA ICE air handling unit model was used, implementing a heat recovery exchanger with 50% efficiency and supply air heating and cooling. Constant supply air temperature was set at 17 °C and fan operation schedule to 6:00–19:00 on workdays. Office and learning spaces were configured for temperature-controlled variable air flows from 2 L/(s m^2) to 4 L/(s m^2), and standard spaces with constant air flows of 2 L/(s m^2). Occupancy, internal load, and indoor lighting schedule was set to 7:00–19:00 on workdays. ASHRAE IWEC2 weather data [33] for Lahti, Finland, and the default IDA ICE urban wind profile were used. The model construction, visible in Figure 1, was simplified by combining zones and windows to ensure practical simulation times with satisfactory resolution. The simplified model contains 111 window and 70 opening or door elements.

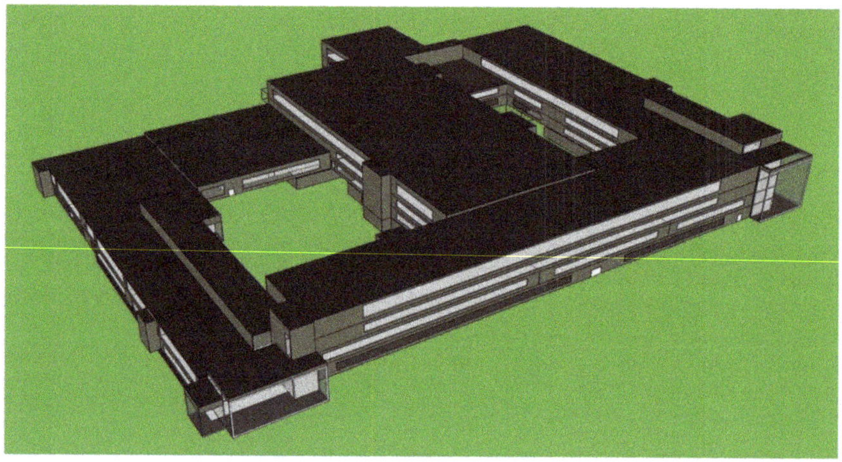

Figure 1. Dynamic campus building energy model constructed in IDA ICE. Walls, roofs, doors, and windows are visible in the three-dimensional representation. The long facade in the forefront is facing west.

The model was validated by comparing monthly simulated space heating and cooling energy consumptions with measured data from the building, which includes GSHP heating and cooling and imported district heat energies for the year 2020. In Finland, annual variations of space heating energy consumption data are standardized according to [34]:

$$Q_{\mathrm{std}} = \frac{S_{\mathrm{N}}}{S_{\mathrm{R}}} Q_{\mathrm{R}} + Q_{\mathrm{DHW}} \tag{1}$$

where Q_{std} is the standardized space heating energy consumption, S_{N} the average reference value for the location, S_{R} the realized reference number for the location, Q_{R} the realized energy consumption for the current year, and Q_{DHW} the domestic hot water energy consumption to be excluded from the standardization. Since the available measured data are consolidated with no possibility to separate the energy used for DHW heating, the correction was performed inversely on the simulated data to convert them to the year 2020. Therefore, Q_{R} is solved by Equation (1), resulting in:

$$Q_{\mathrm{R}} = \frac{S_{\mathrm{R}}}{S_{\mathrm{N}}} (Q_{\mathrm{std}} - Q_{\mathrm{DHW}}), \tag{2}$$

where Q_{std} is now the energy consumption input from the simulated data.

The relative reference values for Lahti, Finland for the whole year of 2020 are $S_R = 3438$ and $S_N = 4392$ [35], indicating a warmer year than usual. In the model validation, the monthly heating degree days values for the year 2020 [35] were used. Monthly comparison of measured energy consumption data and corrected simulated values for model validity assessment is presented in Table 2.

Table 2. Corrected simulated monthly energy consumption data from IDA ICE is compared to the measured data from 2020. The simulated values are corrected to represent the year 2020 through the heating degree days correction. The data are used for the validation of the energy consumption model.

Month	Measured [MWh]	Simulated (IDA ICE) [MWh]	Difference [MWh]
January	1002	899	−103
February	939	835	−104
March	927	776	−150
April	603	671	69
May	410	409	−1
June	246	133	−113
July	249	247	−2
August	223	203	−21
September	274	194	−80
October	465	480	15
November	672	735	64
December	750	890	140
Whole year	6760	6474	−285

Although some monthly and seasonal variance is visible in the data in Table 2, the annual difference of −4.2% and the similar trend shapes between measured and simulated data were considered to validate the energy consumption model for use in this study.

2.2. Modular Energy System Model

The building energy system model was constructed with an open-source linear programming energy system modeling framework Calliope [29]. The framework enables multi-objective optimization with a linear programming solver of choice; in this case, COIN-OR Branch-and-Cut (CBC) solver was used. The problem was defined as linear programming (LP) instead of mixed integer linear programming (MILP), since values such as storage capacity can reduce to zero to represent an absent equipment, and no binary investment costs are included.

The model is based on energy balances of each modeled component, fulfilling the energy balance at each timestep for each separate location, technology, and energy carrier. In this single-building case, there is only one location, and the balance is written as follows [36]:

$$E_{prod}(tech :: carrier) + E_{\ con}(tech :: carrier) + E_{export}(tech :: carrier) = 0 \quad (3)$$

where for a specific technology in an energy carrier ($tech :: carrier$), E_{prod} is the energy produced in or brought to the node, $E_{\ con}$ is consumed, and E_{export} is exported.

In this case, three main energy carriers were included in the model: electricity, heating, and cooling. The heat carrier is divided into high- and low-temperature subcarriers: high-temperature heat is always imported from district heat, while low-temperature heat can be sourced from GSHP or storage. At least 50% of DHW heating power is always sourced from the high-temperature subcarrier. The overall structure of the model components and carriers is presented in Figure 2.

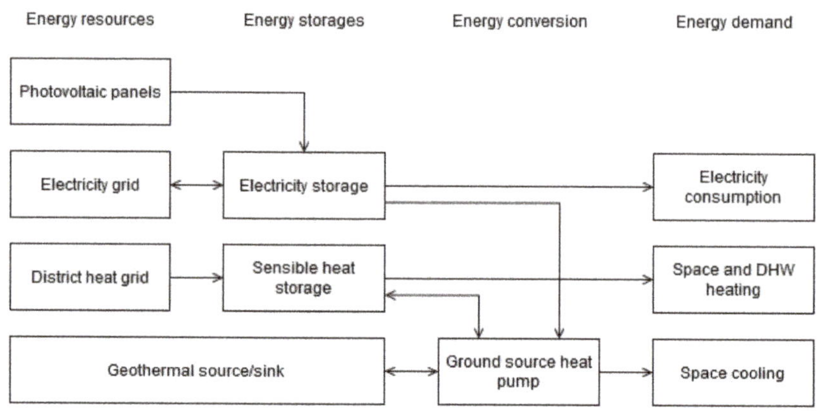

Figure 2. Diagram of the energy system optimization model for this specific case. The main component enabling flexibility is the ground source heat pump, which links separate energy carriers through its energy conversion ability. In this model, energy demands are pre-defined from the dynamic building model data.

The model configuration optimizes three parameters: PV panel peak power, heat storage capacities, and electricity storage capacities. The system's lifetime is 25 years and the real interest rate is 5%. Emissions are considered with no depreciation rate. The following subsubsections describe the individual components modeled in the energy system model, as well as their mathematical formulations.

2.2.1. GSHP Model

The GSHP is a central component in the model, as it enables power-to-heat conversion when excess PV production is available. It was simplified as an energy conversion component with constant conversion ratios, following Equation (4) [36] in each timestep. Using the constant coefficient of performance (COP) for heating and energy efficiency ratio (EER) for cooling is reasonable, as the heat pump evaporator and condenser temperatures stay quite constant in operation.

$$\sum_{tech::carrier,out} \frac{E_{\text{prod}}(tech :: carrier)}{R(tech :: carrier, \text{out})} = -\sum_{tech::carrier,\text{in}} E_{\text{con}}(tech :: carrier) \times R(tech :: carrier, \text{in}) \times \eta(tech) \quad (4)$$

where R is the set carrier conversion ratio. Here, the ratios correspond to COP and EER and include the conversion efficiency $\eta(tech)$. Thus, this specific GSHP is modeled as:

$$\frac{E_{\text{prod}}(\text{GSHP} :: \text{heat})}{\text{COP}} + \frac{E_{\text{prod}}(\text{GSHP} :: \text{cooling})}{\text{EER}} = -E_{\text{con}}(\text{GSHP} :: \text{electricity}) \times 1 \quad (5)$$

The measured GSHP production data from the site data showed a combined production-weighted average COP/EER 3.8, calibrated IDA ICE GSHP model COP 3.3 and EER 4.8, and system datasheets COP 3.3 and EER 4.3 at design operating points. The datasheet values were selected and resulted in approximately 50 °C condensing temperature both in the IDA ICE model and datasheets.

In GSHP systems, too intense heat removal can lead to freezing of borehole water and the surrounding soil [37], and therefore, the borehole heat extraction power must be limited. Real systems are often controlled with pre-set operation modes for different ambient temperatures. In the location, geothermal potential is approximately 7 kW per one 300 m borehole [38], which scales to 368 kW for the entire borehole field. Maximum GSHP compressor power of 300 kW fulfilled the condition on annual level.

2.2.2. Solar Generation and Energy Storage Technologies

The modeled PV capacity was constrained to a range from the present 111 kW to the maximum that was estimated to fit the building roof, i.e., 555 kW. Widely commercially available mono- or polycrystalline silicon panels were chosen, for which price and carbon dioxide-equivalent (CO_2-eq) life cycle emission data are widely available. The used parameters for the PV production component are listed in Table 3.

Table 3. Boundaries and specific parameter for photovoltaic systems.

Parameter	Value
Maximum total peak power [kW_{peak}]	555
Specific cost [€/kW_{peak}]	1000 [39]
Specific emission [kg CO_2-eq/kW_{peak}]	1200 [40,41]

Thermal storage was modeled as a water tank sensible heat storage, as the other alternative, phase change material (PCM) latent heat storage, is not yet technologically mature enough. Furthermore, the available cost and life cycle emission data of PCM storages is scarce and uncertain. Electrical storage was modeled as a lithium-ion battery due to its wide commercial availability and data availability. The simulated storage energy states are defined according to Equation (6) [36], in which the current charge state is defined via the previous timestep state:

$$E_{\text{storage}}(tech, timestep) = E_{\text{storage}}(tech, timestep_{-1}) - E_{\text{con}}\eta(tech) - \frac{E_{\text{prod}}}{\eta(tech)} \quad (6)$$

where $E_{storage}$ is the charge of the observed storage and $\eta(tech)$ is the technological charge and discharge efficiency including all losses. Relevant constraints and values for the thermal and electrical storage systems are listed in Table 4.

Table 4. Cost, emission, and performance parameters for on-site storage components.

Parameter	Thermal Storage	Electricity Storage
Storage capacity limit [MWh]	10	10
Power-to-energy ratio [kW/kWh]	0.1	1
Round trip efficiency [%]	81 [42]	90 [43]
Specific cost [€/kWh]	20 [42]	300 [43]
Specific emission [kg CO_2-eq/kWh]	0.011 [44]	150 [45,46]

2.2.3. Cost and Emission Factors of Energy Grid Connections

Emission factors for imported energies can be defined by accounted, average, marginal. or other values. Although electricity and district heat can be bought as accounted carbon-free products from the utilities, the actual delivered energy can be from mixed sources. In this model, average values for the connected grids were used. The current national electricity emission factor is approximately 72 g CO_2/kWh [47], and the local district heat emission factor is 57 g CO_2/kWh [48]. Emission factors for electricity and district heat are assumed to decrease linearly and reach zero after 25 years. Energy import costs are the total empirical values for the case, including energy, transmission, and tax components. Exported electricity compensation includes only the energy component, and thus, it is lower than the total cost of imported energy. The values are consolidated in Table 5.

Table 5. Cost and emission parameters for energy grid connections.

Parameter	Electricity	District Heat
Specific energy cost [€/kWh]	0.10 [49,50]	0.05 [51]
Exported energy compensation [€/kWh]	0.03 [52]	–
Current CO_2 emission [g CO_2/kWh]	72 [47]	57 [48]
Projection, after 25 years [g CO_2/kWh]	0	0
Averaged over lifetime [g CO_2/kWh]	36	28.5

2.2.4. Life Cycle Cost and Emission Calculation

Life cycle costs and emissions were calculated with Calliope, following Equation (7) [36]. Emissions from maintenance operations were not considered separately, as they are not a significant part of the total life cycle emissions in this case. However, they can be included in other models or applications, if necessary.

$$C(cost, tech) = C_{\text{inv}}(cost, tech) + \sum_{timestep \in timesteps} C_{\text{var}}(cost, tech, timestep) \quad (7)$$

where $C(cost, tech)$ represents the cost for a single cost type ($cost$, such as monetary or CO_2-eq) and for a single technology ($tech$). C_{inv} is the present value of investment and C_{var} is the present value of variable costs for each considered timestep. The total system cost or emission for the entire lifetime is calculated by summing all individual technology cost values calculated with Equation (7).

3. Results

The photovoltaic panel, thermal storage, and electrical storage retrofits were optimally dimensioned for the case building for three objective scenarios, which minimize the objective functions for the system lifetime of 25 years:

- Economic: cost 100%.
- Balanced: cost 50%, emissions 50%.
- Environmental: emissions 100%.

The optimization results are presented in Table 6. Compared to the reference scenario, in all modified scenarios, the GSHP can be at least 5% better utilized in heating due to more available renewable electricity or the added thermal storage component. Electricity storage (lithium-ion battery) is not deployed on any objective due to high life cycle costs and emissions.

As the data in Table 6 shows, in the economic scenario, maximum possible PV generation capacity and some thermal storage is installed. This setup provides a total energy import reduction of over 5% annually, but considering life cycle emissions from the equipment, it causes 6.5 tons more annual CO_2-eq emissions compared to the reference scenario. This converts to a 2.6% life cycle increase. The large increase of PV production capacity leads to more electricity exports than in the reference scenario, which is especially noticeable on the self-used PV electricity ratio of 90% (approximately nine percentage points lower than in the reference results). The rather small compensation of exported electricity 0.03 €/kWh keeps the arrangement profitable.

The balanced scenario increases annual costs slightly while reducing emissions roughly by 1%, enabled by thermal energy storage at 2.5 MWh capacity. It is notable that high CO_2-eq emissions from PV panel production block additional installations even in this half-environmental objective. DH imports are replaced with electricity imports due to more extensive use of GSHP heating.

In the environmental scenario, the thermal storage capacity is increased to the cap of 10 MWh, with an annual cost increase of almost 10 k€. This option seems the least desirable for the building owner, even though the GSHP can be very well utilized.

Table 6. Energy system optimization results and parameters for the three optimization objectives simulated in Calliope, compared to the reference case.

	Reference	Economic	Balanced	Environmental
Installed PV peak power [kW]	111	555	111	111
Installed heat storage [kWh]	0	1982	2619	10,000
Installed electricity storage [kWh]	0	0	0	0
GSHP heat production [TWh/a]	4.92	5.17	5.21	5.39
Production change to reference [%]		+5.0	+5.8	+9.5
PV exports [MWh/a]	0.98	46.05	0.00	0.00
PV self-use ratio [%]	99.0	90.3	100.0	100.0
Electricity import [TWh/a]	5.38	5.12	5.46	5.52
DH import [TWh/a]	2.54	2.35	2.31	2.16
Total energy import [TWh/a]	7.92	7.47	7.78	7.68
Import change to reference [%]		−5.7	−1.8	−3.0
Cost of energy imports [k€/a]	665.1	629.5	662.1	660.2
Cost of investments [k€/a]	0.0	33.1	3.7	14.2
Cost change to reference [%]		−0.4	+0.1	+1.4
CO_2 from energy imports [t CO_2/a]	266.1	251.3	262.6	260.4
CO_2-eq from equipment [t CO_2/a]	0.0	21.3	0.0	0.0
CO_2.-eq change to reference [%]		+2.6	−1.2	−1.9

3.1. *Sensitivity from Imported Energy Prices and Emissions*

Result sensitivity was assessed by varying the uncertain inputs and assumptions of the energy system model: future imported energy price, grid emission factors and their projections. The assessment was performed on three scenarios, simulated for each optimization objective to enable evaluating effects from both monetary and emission factors:

A. All imported electricity and district heat is considered zero-emission for the whole simulation period. Procurement of carbon-neutral energies raise the specific costs of imported energy by 20% in the model. It is acknowledged that full life cycle emissions are neglected, and actual emissions could be compensated in accounting.

B. Instead of carbon-neutral electricity and DH networks after 25 years, the specific emissions will only reduce linearly to 50% from current levels. In the model, the emission factors for imported energy are increased by 50%.

C. This modification assumes that the cost of imported energy will increase linearly by 100% in 25 years, instead of staying at current inflation-corrected levels. In the model, the imported energy prices are increased by 50%.

Figure 3 shows simulation results from the listed scenarios. Overall, the optimization results are very sensitive to assumptions about future energy prices and emissions. Lithium-ion battery storage capacity remains at zero even in all the above scenarios.

The most apparent remark from Figure 3 is that in scenario A with environmental optimization objective, the optimized system obviously prefers importing all energy from the networks instead of installing any local production or storage equipment. This is clear, as installing new local equipment would produce life cycle emissions. However, it well represents the discrepancy of the available carbon emission values and the possibilities of creative CO_2 accounting.

In scenarios B and C, the modifications cause the largest effect in the balanced optimization objective (50% monetary, 50% emission), in which the installed PV power and heat storage capacities are greatly increased due to less desirable imports. However, even in the environmental objective in scenario B, the PV power capacity is not increased due to its high specific emissions.

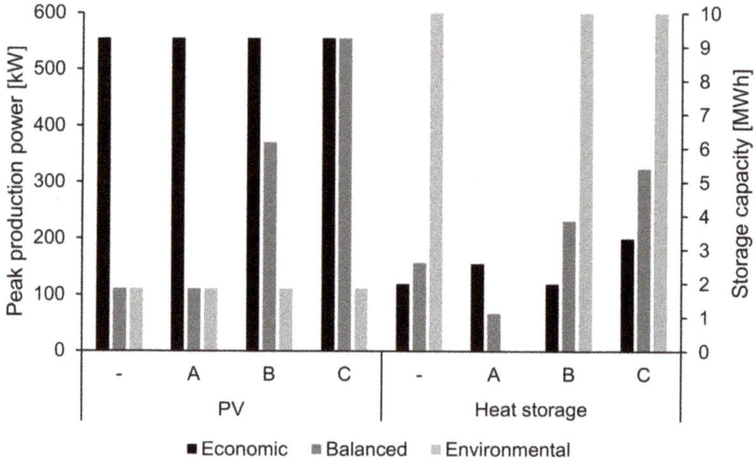

Figure 3. Sensitivity assessment results of optimization, in which the descriptions are as presented earlier. -: No modifications, original results. A: All imported energy is carbon neutral. B: Emission factors are increased 50%. C: Cost factors are increased 50%. Lithium-ion battery capacity remains zero in all scenarios, and thus, it is excluded from the graph.

3.2. *Feasibility of Component Life Cycle Emissions*

As no additional PV or electricity storage capacity were introduced in the environmental optimization results, a feasibility study for component life cycle emissions was conducted to determine the magnitude of necessary life cycle emission reductions to make the component installations environmentally feasible. At the same time, the specific component life cycle emission values sourced from literature are a major source of result uncertainty. Specific emission values for lithium-ion batteries vary greatly in the literature and depend on the manufacturing location. The current values range approximately 80–250 kg CO_2-eq/kWh [45,46], and similarly, broad ranges are present in the PV and heat storage values. Therefore, sensitivity from the life cycle emission factors was assessed by varying the emission factors of all components, as in Table 7. The optimization runs were performed for the environmental optimization objective, as the modifications cause no monetary impacts.

Table 7. Parameters for assessing the effects of varying component life cycle emission values.

Parameter	−75%	−50%	−25%	Initial	+50%
PV panel [kg CO_2-eq/kW_{peak}]	300	600	900	1200	1800
Water tank [kg CO_2-eq/kWh]	0.00275	0.0055	0.00825	0.0110	0.0165
Lithium-ion battery [kg CO_2-eq/kWh]	37.5	75	112.5	150	225

The simulations resulted in configurations presented in Table 8. In this specific case, a lithium-ion battery becomes environmentally feasible only in the −75% scenario, with a specific life cycle emission of 37.5 kg CO_2-eq/kWh. According to the literature, this level is not currently achievable. PV installation becomes feasible in the −50% scenario, resulting in maximum possible installation. The analysis also shows that correct component life cycle emission values are essential for reliable environmental optimization results.

Table 8. Results of the component life cycle emission feasibility analysis.

Parameter	−75%	−50%	−25%	Initial	+50%
PV panel [kW_{peak}]	555	555	111	111	111
Water tank [kWh]	10,000	10,000	10,000	10,000	10,000
Lithium-ion battery [kWh]	51	0	0	0	0

3.3. Power-to-Heat and Storage Interoperation

As heat production with GSHP is cheaper and causes lower emissions than DH even with imported electricity due to the high COP, the model utilizes the GSHP to proactively store heat in the thermal storage for space and DHW heating. This power-to-heat operation mode is especially beneficial in the timesteps where PV production exceeds local electricity demand, which is often the case on weekends. For assessment, the economic scenario optimization result was used because of the maximum available PV generation. The optimized storage capacities are not suitable for seasonal storage, but sufficient for inter-day buffer. Figure 4 shows system operation for the first week (Monday–Friday) of July including the leading weekend (Saturday–Sunday), when excess PV production is available.

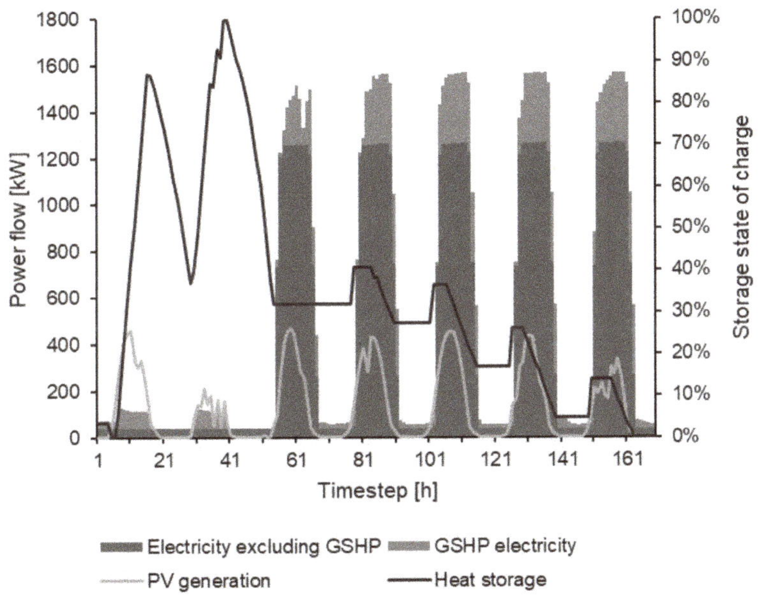

Figure 4. Demonstration of power-to-heat and storage interoperation from the simulated data. The graph shows one-hour timesteps for the first week (Monday–Friday) of July, including the leading weekend (Saturday–Sunday).

Figure 4 shows the thermal storage charging to 100% state-of-charge (SOC) during the weekend using heat generated with GSHP, when the local photovoltaic generation exceeds the total electricity demand. The storage is then optimally discharged during the week by utilizing the heat mainly for DHW heating.

4. Discussion and Conclusions

In this study, a new building energy system optimization approach was developed. The approach was tested for a case campus building, for which energy consumption data were generated with a dynamic energy model. The consumption data were used to develop an energy system optimization model, which in this case was used to dimension photovoltaic (PV) generation and storage equipment. The optimization objectives were both monetary cost and carbon dioxide equivalent (CO_2-eq) emissions. The developed method was identified to function satisfactorily in the case study, and the optimization results served as the material for the feasibility analysis of multi-objective optimization.

The developed energy system optimization model sees input data from dynamic building simulation or measurements as "black box", taken as provided. The black box nature of the input data reduces the computational load and sensitivity in the optimization phase. For existing buildings, measured data could also be used. The approach prevents considering energy demand flexibility, and possible thermodynamic effects to the building structures are also neglected. Comprehensive modeling of these factors would require a more integrated optimization solution, potentially implemented directly in the dynamic building simulation software. Integrated optimization, however, requires considerably more computational resources, while the industry needs an easily applicable and computationally affordable approach. The developed method fulfills this requirement.

The power-to-heat concept manifested naturally from the available component mix of PV panels, thermal storage, and heat pump. In this case, it was demonstrated as the capability to convert excess solar electricity production to heat with the heat pump and store it in the thermal storage for later consumption. This concept could turn out as an essential function in future carbon-free energy systems. Studying this approach more specifically, however, is outside the scope of this study and requires further research.

The retrofit dimensioning results of the case campus building are significant and puzzling. Due to high emissions from manufacturing of photovoltaic panels and lithium-ion batteries in addition to low CO_2 factors of imported energy, no life cycle emission reductions appear to be gained from installing additional PV panels or batteries for the 25-year lifetime in this case. This result significantly diverges from the presumption that local renewable energy production and storage provide environmental life cycle emission benefits. The conclusion is not directly generalizable, however, as it depends on the specific building and its equipment, local CO_2 emission factors, and projections that electricity and district heat imported from the energy networks will be carbon neutral after 25 years. Future research needs to cover more buildings in different locations to study differences between geographical areas.

The sensitivity and feasibility analyses of the case results indicate that correct imported energy and component life cycle emission values are crucial for reliable environmental optimization. Sourcing of correct life cycle emission data for components such as lithium-ion batteries requires knowledge of the manufacturing location and detailed specifications. Furthermore, emission factors of imported grid energy are rapidly decreasing, as electricity and DH networks in Finland are converting to carbon-neutral production, and the reported values do not always represent the full life cycle. This complicates the sourcing of reliable input values for environmental optimization and causes the results to expire rapidly.

The presented optimization approach considers only a single building. For cost optimization, this seems feasible, as the incentives are clear for the property owner in the form of cost savings. However, emission minimization has currently no tangible incentive for the property owner and may thus remain unimplemented. Furthermore, the environmental optimization result with no installed PV panels seems to conflict with the fact that small-scale distributed renewable energy generation including PV is generally considered a requirement for carbon-neutral electricity networks. Applicable results would probably require optimization on several system levels and coordination between them. In the future, frameworks such as emissions trade could be further extended to translate these considerations into the monetary domain.

Based on the previous points, it would seem feasible that especially in future renewable energy systems, reported specific energy emissions included life cycle emissions from system manufacture, deployment, operation, decommissioning, and other considerations, instead of only direct emissions from combustion. The current methodological discrepancies in emission factors complicate emission optimization and cause uncertainty in the results. Naturally, the situation might change with future regulations or incentives, but the field still needs further research. Currently, for a building owner, single-objective cost optimization could be the most sensible approach due to its clear incentives and robustness.

Author Contributions: Conceptualization, V.E., A.G. and A.W.; data curation, V.E. and A.W.; investigation, V.E., A.G. and A.W.; methodology, V.E. and A.G.; project administration, A.G.; software, V.E.; validation, V.E. and A.G.; writing—original draft, V.E. and A.W.; writing—review and editing, V.E., A.G. and A.W. All authors have read and agreed to the published version of the manuscript.

Funding: This study received funding from the European Regional Development Fund (ERDF) under project A75174.

Institutional Review Board Statement: Not applicable.

Informed Consent Statement: Not applicable.

Data Availability Statement: The data presented in this study are available on request from the corresponding author.

Conflicts of Interest: The authors declare no conflict of interest. The funders had no role in the design of the study; in the collection, analyses, or interpretation of data; in the writing of the manuscript, or in the decision to publish the results.

Appendix A

Appendix A contains floor plans of the modeled case building. Variable z represents the vertical coordinate.

Figure A1. First floor (z = 0.0 m).

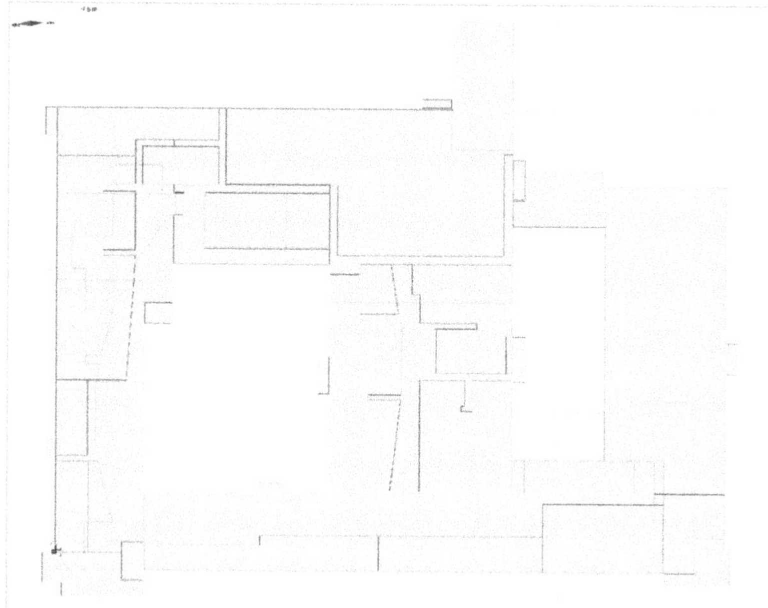

Figure A2. Second floor (z = 4.5 m).

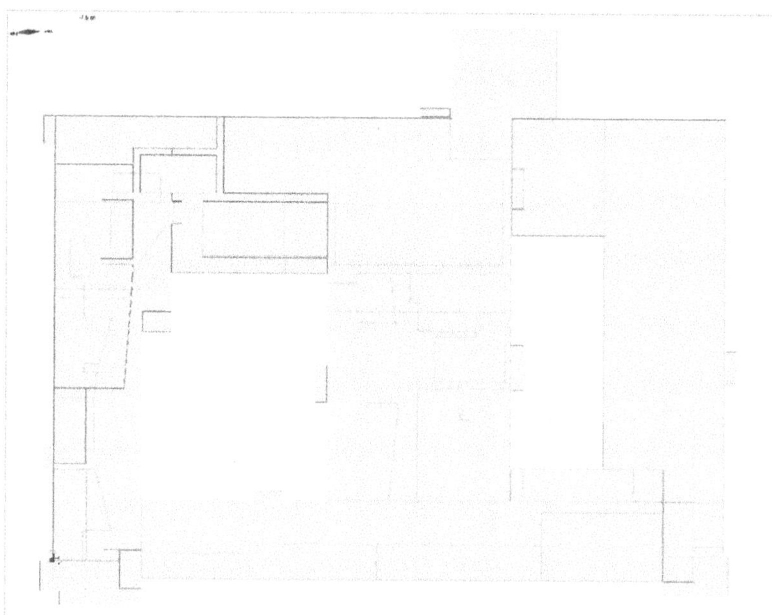

Figure A3. Third floor (z = 9.0 m).

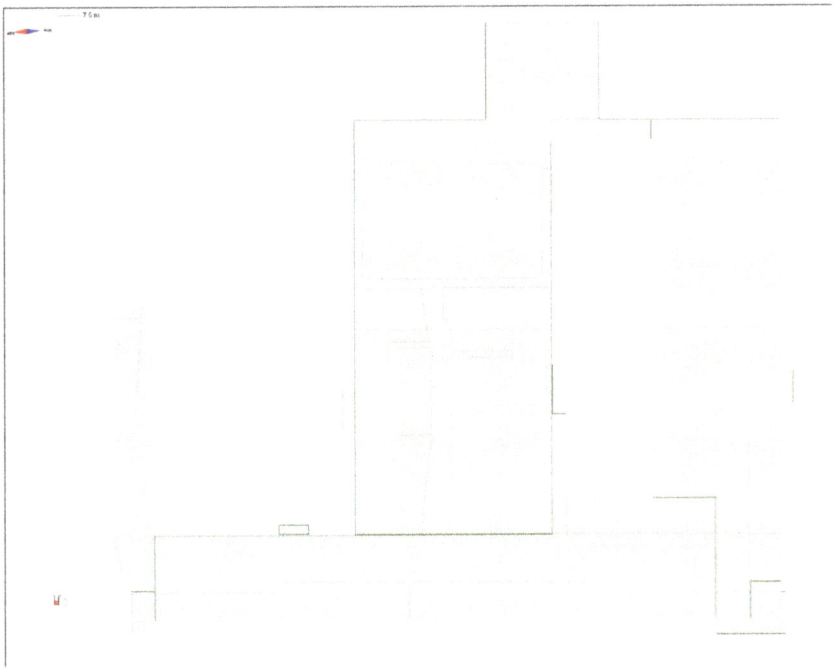

Figure A4. Fourth floor (z = 13.5 m).

References

1. IEA. Buildings–Topics. Available online: https://www.iea.org/topics/buildings (accessed on 16 June 2021).
2. European Commission. Energy Performance of Buildings Directive. Available online: https://ec.europa.eu/energy/topics/energy-efficiency/energy-efficient-buildings/energy-performance-buildings-directive_en (accessed on 16 June 2021).
3. Motiva. Aurinkosähköjärjestelmän Mitoitus (Solar Electricity System Dimensioning). Available online: https://www.motiva.fi/ratkaisut/uusiutuva_energia/aurinkosahko/hankinta_ja_asennus/aurinkosahkojarjestelman_mitoitus (accessed on 16 June 2021).
4. EQUA Simulation AB. IDA ICE-Simulation Software. Available online: https://www.equa.se/en/ida-ice (accessed on 18 June 2021).
5. National Renewable Energy Laboratory; U.S. Department of Energy. EnergyPlus. Available online: https://energyplus.net/ (accessed on 30 June 2021).
6. Thermal Energy System Specialists, LLC. TRNSYS: Transient System Simulation Tool. Available online: http://www.trnsys.com/ (accessed on 30 June 2021).
7. Becchio, C.; Ferrando, D.G.; Fregonara, E.; Milani, N.; Quercia, C.; Serra, V. The Cost-Optimal Methodology for the Energy Retrofit of an Ex-Industrial Building Located in Northern Italy. *Energy Build.* **2016**, *127*, 590–602. [CrossRef]
8. Wu, R.; Mavromatidis, G.; Orehounig, K.; Carmeliet, J. Multiobjective Optimisation of Energy Systems and Building Envelope Retrofit in a Residential Community. *Appl. Energy* **2017**, *190*, 634–649. [CrossRef]
9. Ascione, F.; Bianco, N.; De Stasio, C.; Mauro, G.M.; Vanoli, G.P. A New Methodology for Cost-Optimal Analysis by Means of the Multi-Objective Optimization of Building Energy Performance. *Energy Build.* **2015**, *88*, 78–90. [CrossRef]
10. Bingham, R.D.; Agelin-Chaab, M.; Rosen, M.A. Whole Building Optimization of a Residential Home with PV and Battery Storage in The Bahamas. *Renew. Energy* **2019**, *132*, 1088–1103. [CrossRef]
11. Delgarm, N.; Sajadi, B.; Kowsary, F.; Delgarm, S. Multi-Objective Optimization of the Building Energy Performance: A Simulation-Based Approach by Means of Particle Swarm Optimization (PSO). *Appl. Energy* **2016**, *170*, 293–303. [CrossRef]
12. Hamdy, M.; Mauro, G.M. Multi-Objective Optimization of Building Energy Design to Reconcile Collective and Private Perspectives: CO_2-Eq vs. Discounted Payback Time. *Energies* **2017**, *10*, 1016. [CrossRef]
13. Milan, C.; Bojesen, C.; Nielsen, M.P. A Cost Optimization Model for 100% Renewable Residential Energy Supply Systems. *Energy* **2012**, *48*, 118–127. [CrossRef]
14. Privitera, G.; Day, A.R.; Dhesi, G.; Long, D. Optimising the Installation Costs of Renewable Energy Technologies in Buildings: A Linear Programming Approach. *Energy Build.* **2011**, *43*, 838–843. [CrossRef]
15. Di Pilla, L.; Desogus, G.; Mura, S.; Ricciu, R.; Di Francesco, M. Optimizing the Distribution of Italian Building Energy Retrofit Incentives with Linear Programming. *Energy Build.* **2016**, *112*, 21–27. [CrossRef]

16. Lindberg, K.B.; Doorman, G.; Fischer, D.; Korpås, M.; Ånestad, A.; Sartori, I. Methodology for Optimal Energy System Design of Zero Energy Buildings Using Mixed-Integer Linear Programming. *Energy Build.* **2016**, *127*, 194–205. [CrossRef]
17. Omu, A.; Choudhary, R.; Boies, A. Distributed Energy Resource System Optimisation Using Mixed Integer Linear Programming. *Energy Policy* **2013**, *61*, 249–266. [CrossRef]
18. Ren, F.; Wei, Z.; Zhai, X. Multi-Objective Optimization and Evaluation of Hybrid CCHP Systems for Different Building Types. *Energy* **2021**, *215*, 119096. [CrossRef]
19. Xu, Y.; Zhang, G.; Yan, C.; Wang, G.; Jiang, Y.; Zhao, K. A Two-Stage Multi-Objective Optimization Method for Envelope and Energy Generation Systems of Primary and Secondary School Teaching Buildings in China. *Build. Environ.* **2021**, *204*, 108142. [CrossRef]
20. Attia, S.; Hamdy, M.; O'Brien, W.; Carlucci, S. Assessing Gaps and Needs for Integrating Building Performance Optimization Tools in Net Zero Energy Buildings Design. *Energy Build.* **2013**, *60*, 110–124. [CrossRef]
21. Nguyen, A.-T.; Reiter, S.; Rigo, P. A Review on Simulation-Based Optimization Methods Applied to Building Performance Analysis. *Appl. Energy* **2014**, *113*, 1043–1058. [CrossRef]
22. Li, L. Integrating Climate Change Impact in New Building Design Process: A Review of Building Life Cycle Carbon Emission Assessment Methodologies. *Clean. Eng. Technol.* **2021**, *5*, 100286. [CrossRef]
23. Ibañez Iralde, N.S.; Pascual, J.; Salom, J. Energy Retrofit of Residential Building Clusters. A Literature Review of Crossover Recommended Measures, Policies Instruments and Allocated Funds in Spain. *Energy Build.* **2021**, *252*, 111409. [CrossRef]
24. Galatioto, A.; Ciulla, G.; Ricciu, R. An Overview of Energy Retrofit Actions Feasibility on Italian Historical Buildings. *Energy* **2017**, *137*, 991–1000. [CrossRef]
25. Gagliano, A.; Nocera, F.; Tina, G. Performances and Economic Analysis of Small Photovoltaic–Electricity Energy Storage System for Residential Applications. *Energy Environ.* **2020**, *31*, 155–175. [CrossRef]
26. He, Q.; Hossain, M.U.; Ng, S.T.; Augenbroe, G. Identifying Practical Sustainable Retrofit Measures for Existing High-Rise Residential Buildings in Various Climate Zones through an Integrated Energy-Cost Model. *Renew. Sustain. Energy Rev.* **2021**, *151*, 111578. [CrossRef]
27. Nocera, F.; Giuffrida, S.; Trovato, M.R.; Gagliano, A. Energy and New Economic Approach for Nearly Zero Energy Hotels. *Entropy* **2019**, *21*, 639. [CrossRef] [PubMed]
28. Pirmohamadi, A.; Dastjerdi, S.M.; Ziapour, B.M.; Ahmadi, P.; Rosen, M.A. Integrated Solar Thermal Systems in Smart Optimized Zero Energy Buildings: Energy, Environment and Economic Assessments. *Sustain. Energy Technol. Assess.* **2021**, *48*, 101580. [CrossRef]
29. Pfenninger, S.; Pickering, B. Calliope: A Multi-Scale Energy Systems Modelling Framework. *J. Open Source Softw.* **2018**, *3*, 825. [CrossRef]
30. Child, M.; Breyer, C. Vision and Initial Feasibility Analysis of a Recarbonised Finnish Energy System for 2050. *Renew. Sustain. Energy Rev.* **2016**, *66*, 517–536. [CrossRef]
31. Litjens, G.B.M.A.; Worrell, E.; Van Sark, W.G.J.H.M. Lowering Greenhouse Gas Emissions in the Built Environment by Combining Ground Source Heat Pumps, Photovoltaics and Battery Storage. *Energy Build.* **2018**, *180*, 51–71. [CrossRef]
32. Baeten, B.; Rogiers, F.; Helsen, L. Reduction of Heat Pump Induced Peak Electricity Use and Required Generation Capacity through Thermal Energy Storage and Demand Response. *Appl. Energy* **2017**, *195*, 184–195. [CrossRef]
33. Huang, Y.J.; Su, F.; Seo, D.; Krarti, M. Development of 3012 IWEC2 Weather Files for International Locations (RP-1477). *ASHRAE Trans.* **2014**, *120*, 340–355.
34. Motiva. Kulutuksen Normitus (Standardisation of Consumption). Available online: https://www.motiva.fi/julkinen_sektori/kiinteiston_energiankaytto/kulutuksen_normitus (accessed on 18 June 2021).
35. Finnish Meteorological Institute. Heating Degree Days. Available online: https://en.ilmatieteenlaitos.fi/heating-degree-days (accessed on 19 August 2021).
36. Pfenninger, S.; Pickering, B.; Tröndle, T.; Garchery, M.; Hawker, G.; Lombardi, F.; Hilbers, A. Calliope 0.6.7 documentation. Available online: https://calliope.readthedocs.io/en/v0.6.7/ (accessed on 27 August 2021).
37. Eslami-nejad, P.; Bernier, M. Freezing of Geothermal Borehole Surroundings: A Numerical and Experimental Assessment with Applications. *Appl. Energy* **2012**, *98*, 333–345. [CrossRef]
38. GTK (Geological Survey of Finland). Maankamara (Earth Crust Map). Available online: https://gtkdata.gtk.fi/Maankamara/ (accessed on 17 June 2021).
39. Philipps, S.; Warmuth, W. Photovoltaics Report. Fraunhofer Institute for Solar Energy Systems. Available online: https://www.ise.fraunhofer.de/content/dam/ise/de/documents/publications/studies/Photovoltaics-Report.pdf (accessed on 5 August 2021).
40. Yıldız, G.; Çalış, B.; Gürel, A.E.; Ceylan, İ. Investigation of Life Cycle CO_2 Emissions of the Polycrystalline and Cadmium Telluride PV Panels. *Environ. Nanotechnol. Monit. Manag.* **2020**, *14*, 100343. [CrossRef]
41. Liu, F.; Van den Bergh, J.C.J.M. Differences in CO_2 Emissions of Solar PV Production among Technologies and Regions: Application to China, EU and USA. *Energy Policy* **2020**, *138*, 111234. [CrossRef]
42. Hauer, A.; Simbolotti, G.; Tosato, G.; Gielen, D. *Thermal Energy Storage*; IEA-ETSAP: Paris, France; IRENA: Abu Dhabi, United Arab Emirates, 2013.
43. IRENA. *Electricity Storage and Renewables: Costs and Markets to 2030*; International Renewable Energy Agency: Abu Dhabi, United Arab Emirates, 2017; ISBN 978-92-9260-038-9.

44. Thaker, S.; Oni, A.O.; Gemechu, E.; Kumar, A. Evaluating Energy and Greenhouse Gas Emission Footprints of Thermal Energy Storage Systems for Concentrated Solar Power Applications. *J. Energy Storage* **2019**, *26*, 100992. [CrossRef]
45. Philippot, M.; Alvarez, G.; Ayerbe, E.; Van Mierlo, J.; Messagie, M. Eco-Efficiency of a Lithium-Ion Battery for Electric Vehicles: Influence of Manufacturing Country and Commodity Prices on GHG Emissions and Costs. *Batteries* **2019**, *5*, 23. [CrossRef]
46. Pucker-Singer, J.; Aichberger, C.; Zupančič, J.; Neumann, C.; Bird, D.N.; Jungmeier, G.; Gubina, A.; Tuerk, A. Greenhouse Gas Emissions of Stationary Battery Installations in Two Renewable Energy Projects. *Sustainability* **2021**, *13*, 6330. [CrossRef]
47. Fingrid. Sähköjärjestelmän tila-CO_2 (Electricity System State-CO_2). Available online: https://www.fingrid.fi/sahkomarkkinat/sahkojarjestelman-tila/co2/ (accessed on 21 June 2021).
48. Lahti Energia. Vastuullisuusraportti (Sustainability Report). Available online: https://www.lahtienergia.fi/vastuullisuusraportti/ (accessed on 21 June 2021).
49. Statistics Finland. Energy Prices. ISSN=1799-800X. 1st Quarter 2021, Appendix Figure 5. Price of Electricity by Type of Consumer. Available online: https://www.stat.fi/til/ehi/2021/01/ehi_2021_01_2021-06-10_kuv_005_en.html (accessed on 5 August 2021).
50. LE-Sähköverkko Oy. Verkkopalveluhinnasto 1.8.2019 alkaen (Electricity Network Service Price List as of 2019-08-01). Available online: https://www.lahtienergia.fi/wp-content/uploads/2021/06/verkkopalveluhinnasto_01082019.pdf (accessed on 5 August 2021).
51. Lahti Energia. Lämpömaksuhinnasto (Heat Price List). Available online: https://www.lahtienergia.fi/wp-content/uploads/2021/06/Lampomaksuhinnasto-01122020-ilman-palveluhinnastoa.pdf (accessed on 5 August 2021).
52. Statistics Finland. Energy Prices. ISSN=1799-800X. 1st Quarter 2021, Appendix Figure 6. Average Monthly Spot Prices at the Nord Pool Spot Power Exchange. Available online: https://www.stat.fi/til/ehi/2021/01/ehi_2021_01_2021-06-10_kuv_006_en.html (accessed on 5 August 2021).

energies

MDPI

Article

Accelerating Building Energy Retrofitting with BIM-Enabled BREEAM-NL Assessment

Vyshali Simhachalam [1], Tong Wang [2,*], Yan Liu [1], Hans Wamelink [2], Lorena Montenegro [3] and Geert van Gorp [3]

1 Faculty of Civil Engineering & Geosciences, Delft University of Technology, 2628 CD Delft, The Netherlands; vyshali.simhachalam@deerns.com (V.S.); Y.Liu-9@tudelft.nl (Y.L.)
2 Faculty of Architecture and the Built Environment, Delft University of Technology, 2628 CD Delft, The Netherlands; J.W.F.Wamelink@tudelft.nl
3 Deerns Nederland B.V., 2595 DA Den Haag, The Netherlands; lorena.montenegro@deerns.com (L.M.); geert.van.gorp@deerns.com (G.v.G.)
* Correspondence: t.wang-12@tudelft.nl

Abstract: The Paris Agreement requires building retrofitting practices to be more efficient and effective. However, the current practice for building energy retrofitting is lacking behind, and one reason for that is the time-consuming process of energy credit evaluation. Energy performance assessment such as BREEAM-NL in the Netherlands could apply a more automatic approach with the help of building information modelling (BIM) for an efficient building energy retrofitting evaluation process. However, to what extent BIM can help in accelerating energy performance evaluation in the BREEAM-NL certification process is under-examined. This paper first combines literature findings with practical interviews from a case study organization to present a holistic overview of the potential for automating energy-related credits evaluation in BREEAM-NL using BIM. To understand the possible impacts of such transition, a responsible, accountable, consulted, and informed (RACI) matrix is developed to map the impacts on different actors involved. Furthermore, to help practitioners in an organizational context to adopt a BIM-enabled energy credits assessment workflow, the case study organization is studied to (1) understand their current BIM use status; (2) propose a suitable starting point to take toward a BIM-enabled energy performance assessment for building energy retrofitting. Finally, the proposed starting point is demonstrated using a customized application, and the project team's feedback is used to verify its efficiency and future directions are identified.

Keywords: building energy retrofitting; building information modelling (BIM); energy performance evaluation; BREEAM-NL; energy transition; RACI matrix

Citation: Simhachalam, V.; Wang, T.; Liu, Y.; Wamelink, H.; Montenegro, L.; van Gorp, G. Accelerating Building Energy Retrofitting with BIM-Enabled BREEAM-NL Assessment. *Energies* **2021**, *14*, 8225. https://doi.org/10.3390/en14248225

Academic Editors: Cynthia Hou and Joseph H. K. Lai

Received: 3 November 2021
Accepted: 3 December 2021
Published: 7 December 2021

1. Introduction

In 2015, 196 different countries signed the Paris agreement to tackle the climate challenges by reducing carbon emissions. In total, the involved countries are responsible for 96.5% of the total emission. This agreement obliges countries to transform their activities on different levels and to be environmentally, socially, and economically ready for the future [1]. The Netherlands signed the Paris Agreement as well, and the conditions stated in the agreement are translated by the Dutch government to a National Climate Agreement (NCA). The central goal of the NCA is to reduce the Netherlands' greenhouse gas emissions by 49% by 2030, compared with 1990 levels, and a 95% reduction by 2050 [2]. The green revolution has urged the architecture, engineering, and construction (AEC) sector's transition, as approximately 35% of the total energy consumption in the Netherlands is accounted for by the building industry [3]. As a result, the Netherlands is on the cusp of a sustainable energy transition of the built environment and the adaptation of the seven million homes and one million buildings [4].

One of the major challenges to achieving sustainable energy transition of the built environment is the state of the existing building stock. Evaluating and optimizing the energy performance of the building is an important part of sustainable building retrofits. However, not knowing the current status and the potential energy benefits in time hinders the implementation of building energy retrofitting. Green building certifications can serve as useful instruments of guidance in this process. Through in-depth evaluation metrics and a decision support framework, green building rating systems can help designers in improving the sustainability of buildings [5].

There are several different environmental certification methods prevalent in the Netherlands such as Building Research Establishment Environmental Assessment Method (BREEAM), Leadership in Energy and Environmental Design (LEED), Enegielabel, GPR Gebouw, etc. Amongst these, BREEAM is one of the most widely accepted choices of environmental assessments [6]. The Dutch Green Building Council (DGBC) has developed a national adaptation of the guideline named BREEAM-NL in 2009. The requirements of the certification system were tailored to align with the national standards and legislation. Ever since, its popularity and the total BREEAM-NL certified area has been increasing steadily. Due to a lack of data, especially for the existing building stock's renovation projects, the sustainability assessments in practice are often conducted at the end of design stages with time-consuming manual data gathering processes, when the scope for changes or corrections is too little [7–10].

The development of building information modelling (BIM) brings the potential to improve this green building evaluation process and accelerate the building energy retrofitting practice. The need for data-driven design has been boosting the application of BIM in the AEC industry. Lately, the use of BIM for the creation of sustainable assets has been gaining momentum. This convergence of BIM and green buildings is often referred to as Green BIM. Krygiel et al. [11] were one of the early works that discussed the potential of BIM for sustainable project delivery. Integration of multidisciplinary information and facilitation of performance analyses with regards to energy, thermal comfort, daylighting, etc., are some of the commonly known advantages BIM can offer for green building projects. Lu et al. [12] identified that BIM-supported lifecycle functions, environmental analyses, and green building assessments are the three primary facets of integrating BIM with green buildings.

Despite the increasing number of studies providing evidence to support these claims, BIM is not being actively used for sustainability assessments or green certifications [13,14]. Furthermore, Ayman et al. [15] concluded that further work must focus on real-life problems identified from the industry and propose solutions in response to that.

The main question that will be answered in this research is, therefore:

> *To what extent and how can BIM help in accelerating building energy retrofitting implementation in the energy performance evaluation process of BREEAM-NL certification practice?*

The research has three main aims:

AIM 1: To investigate in industrial practice, supported by literature, the possibilities of BIM application in building energy retrofit assessments for BREEAM-NL certification;

AIM 2: To understand the impacts on actors involved in the transition to an automated BIM-enabled approach;

AIM 3: To propose context-specific starting points in response so that the building energy performance evaluation can be automated, and the building energy retrofitting process can be accelerated.

The rest of the paper is structured as follows:

Section 2 presents the context of the research, including both literature findings and an introduction to the case study organization;

Section 3 explains the methodology followed in this research;

Section 4 presents the possibilities for BIM to enable automatic energy-related credits evaluation in BREEAM-NL assessment;

Section 5 presents the possible impacts for different actors involved based on a responsibility, accountable, consulted, and informed (RACI) matrix;
Section 6 makes a customized recommendation starting point for this case study organization based on their BIM maturity and workflow and the recommendation is demonstrated using a custom application for one selected credit. The impacts are validated by experts;
Section 7 concludes the findings and discusses the limitations and future directions.

2. Context of the Research

2.1. BREEAM-NL for Building Energy Retrofits

The world is witnessing a paradigm shift leading to the evolution of building codes and guidelines to enable and evaluate sustainable developments, such as the development of green building councils [16], which promote green building assessments (GBA). The Building Research Establishment's Environmental Assessment Method (BREEAM) has become more popular in European countries [17] due to its flexibility for adaptation by the local regulations. BREEAM-NL is the national adaptation of this international scheme tailored to the regional context of the Netherlands. An important distinction has been made in the roles involved in the accreditation process. BREEAM international only has a licensed assessor that is responsible for assessing the documentation submitted by the design team, while BREEAM-NL has two functional roles: BREEAM-NL expert and BREEAM-NL assessor. A BREEAM-NL expert is a trained content and process manager that can support the developer/ the client in the design and construction phases to meet the BREEAM-NL requirements. A BREEAM-NL assessor is an independent professional working for a licensed organization. The assessor is responsible for examining the evidence submitted by the expert and preparing an assessment report based on which DGBC issues the final certification decision.

The BREEAM-NL assessment starts with the client's ambition, expressed as a program of requirements, which guides the design process for architects and mechanical, electrical, and plumbing (MEP) designers. The design teams provide information to the BREEAM-NL team that will be used to verify the compliance of the design with the certification requirements. The interaction and feedback will be used to optimize the design. The BREEAM-NL team submits the documentation to an assessor. The chaotic distribution of information is one of the biggest challenges of green building assessments (GBA) [18]. This manual way of green building assessments is not only time consuming but also error prone [19]. A more efficient and intelligent way of data acquisition and assessment system can greatly benefit project teams in achieving green certifications.

BREEAM-NL rating system has three hallmarks depending on the type of the project: new construction and renovation, in use, area development and re-development. Large-scale renovations or deep energy retrofits that involve changes to the building envelope, installations, or the building function fall under the first hallmark: new construction and renovation. Energy is one of the ten sustainability topics covered in the BREEAM-NL guideline. The requirements related to this category weigh the highest, amounting to 19% of the total available points [20].

To obtain a BREEAM-NL certificate for a renovation project, the project teams must demonstrate compliance with numerous requirements laid out in the assessment guideline. These requirements concern improvement in the performance of the asset as compared with the existing baseline situation. Additionally, based on the total number of points obtained, the certification decision is made. This entire process is time consuming and laborious, because it requires large amounts of interdisciplinary information distributed between several different stakeholders. Moreover, the availability of accurate information about the existing building condition and its performance during the design phase is crucial for a successful renovation project.

In the Netherlands, approximately half of the housing stock was built before 1975, and nearly 40% of the usable surface area in non-residential building stock was built after 1994 [21]. Therefore, most of these assets, especially in the non-residential sector, are not energy efficient and do not yet have a registered energy label. Energy renovation of existing buildings is instrumental to meeting future sustainable requirements [22]. As mentioned, BREEAM-NL assessment can help make optimal designs through energy performance metrics, and automating this evaluation process can improve the building energy retrofitting practice.

2.2. BIM-Enabled Energy Performance Assessments

BIM refers to the ICT technologies, working processes, and policy guidelines that together enable the creation and management of information related to a building throughout its lifecycle [23]. By integrating the different data sources in a single information model, BIM not only facilitates n-dimensional analysis but also allows for smoother, real-time collaboration between project stakeholders. In light of these advantages, the majority of the AEC organizations started to move away from two-dimensional CAD workflows toward a data-oriented, BIM approach in the past two decades. This transition occurs in incremental maturity stages starting from object-based 3D modelling to network-based integration of all disciplines [24].

Energy performance assessment and compliance verification is one of the many opportunities created by the introduction of BIM. For example, through building energy models (BEM), BIM can enable the assessment of the performance of an asset against energy efficiency standards. It can also help identify the best renovation and retrofitting solutions by allowing for a comparison between existing building performance and the expected performance after selected changes [25]. By enabling performance evaluation in concurrence with the design process, it empowers design teams to make sustainable choices in the early design phases of a project when the flexibility for changes and corrections is the highest. The use of BIM also offers the possibility of automating the compliance verification process for green certifications, which greatly reduces the time and errors associated with traditional methods of sustainability assessments. Storing information related to green certification requirements in the BIM model could also fasten the process of producing necessary documentation. The use of BIM can, therefore, benefit the energy assessment process in three ways:

- By enabling performance evaluation in concurrence with the design process, it empowers design teams to make sustainable choices in the early stages of the project;
- By replacing the traditional, manual methods of evaluation with an automated approach, it makes the process of verifying compliance with regulations or green certification requirements more efficiently;
- By proper data management and linking information sources, it facilitates and streamlines the massive document management required for achieving these certifications [12,26,27].

Despite the increasing number of studies providing evidence to support this claim, the industry is struggling to capitalize on the benefits of green BIM. As of now, Lu et al. [12] identified several challenges facing the implementation of BIM-based energy and sustainability assessments such as a lack of appropriate technology, interoperability between BIM and BEM tools, and the integrity of BIM models. Furthermore, most of the studies (more than 70%) on the integration of BIM and sustainability assessments focused on LEED certification, while knowledge regarding BIM applicability for other certifications is insufficient [28]. Based on the existing literature, it is estimated that about 67% of LEED credits can be linked to BIM, whereas merely 24% of BREEAM credits have been linked to BIM [5]. Furthermore, the requirements in BREEAM-NL are aligned with the Dutch legislation, and these are not yet embedded into the commonly used software programs. Therefore, until the release of a relevant commercial software, in order to automate assessments, technical infrastructure has to be developed by the project teams themselves. Table 1 provides an overview of the software tools used for automated assessments of various sustainability rating systems. It can be seen that BREEAM-NL was not discussed before.

Table 1. Common software tools used for sustainability evaluations (derived from Carvalho et al. [5]).

Software	Applied Certifications [1]
Autodesk REVIT (in combination with API/Dynamo)	BREEAM International 2013, BREEAM Offices 2008, LEED NC
IES-VE	BREEAM UK 2011, BREEAM UK Refurbishment and Fit-out 2014, LEED NC
TAS	BREEAM UK 2011, LEED NC
Energy Plus	BREEAM UK 2011, LEED NC
Green Building Studio	LEED, BREEAM UK Refurbishment and Fit-out 2014
Visual Studio	BREEAM Europe Commercial 2009
Excel	BREEAM International 2013, LEED NC

[1] Software tools used in previous academic works for the listed rating systems.

2.3. Energy-Related Credits in BREEAM-NL

The BREEAM-NL assessment guideline is divided into nine "categories" based on the topic they deal with. Figure 1 shows the different categories and their weightage in the new construction and renovation hallmark of BREEAM-NL. Energy category has the highest weightage at 19% followed by health and comfort at 15% and materials at 12%.

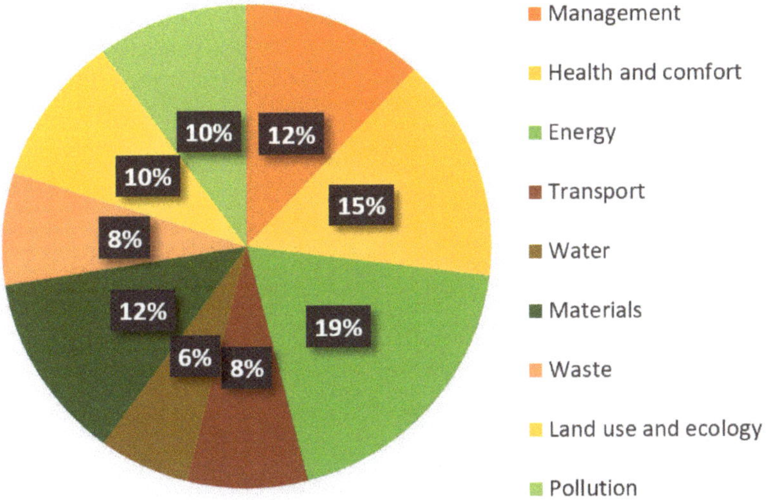

Figure 1. BREEAM-NL new construction and renovation—composition.

The energy category of BREEAM-NL is divided into credits aiming at measures to improve the energy performance of the building through design features and efficient installations. These credits are worth a maximum of 29 points, plus an additional 2 points available for exemplary performance. The description of the credit requirements and associated points are provided in Table 2. This table serves as the input for the semi-structured interviews conducted to investigate the potential of using BIM for automated assessments in Section 4, the associated impacts of BIM use on different actors in Section 5, and recommended digitalization starting points in Section 6.

Table 2. Overview of the credits in the energy category, BREEAM-NL 2014 v2 (derived from the BREEAM-NL guideline).

Credit Description	Max Score	Sustainability Objective	Assessment Method
ENE 01: Energy efficiency	15	To encourage design optimization that will result in the lowest possible CO_2 emissions due to building-related energy consumption	Percentage improvement in energy performance coefficient (EPC) as compared with the energy performance standard (EPN) has to be calculated
Exemplary performance	2	To promote exemplary performance through CO_2 neutral building (parts) and dynamic energy modelling	Calculation of energy generation and demands of the building through dynamic modelling tools
ENE 02: Sub-metering of energy consumption	2	To ensure that the significant energy consumption zones within a building are metered and monitored separately	Design verification to ensure that energy sub-meters are placed in the significant consumption groups
ENE 04: Energy-efficient outdoor lighting	1	To promote the usage of energy-efficient lighting fixtures and reduce outdoor lighting related CO_2 emissions	Specific lighting power per lux calculation and verifying if it is under 0.1 W-Lux/m^2.
ENE 05: Application of renewable energy	3	To encourage the use of renewable energy sources	Feasibility study for the application of renewable energy sources and the resulting percentage reduction in carbon emissions
ENE 06: Minimizing air filtration	1	To promote CO_2 reduction through efficient design that minimizes heat and cold losses	Qualitative design verification to ensure the application of appropriate interventions for minimal loss of heat and cold
ENE 07: Energy-efficient refrigeration and cold storage	1	To promote energy savings and CO_2 reduction through the use of efficient cold storage equipment	Verifying that the specifications of the refrigeration equipment meet the requirements
ENE 08: Energy-efficient elevators	2	To promote energy savings and CO_2 reduction through efficient elevators	Verifying that the specifications of the elevators meet the requirements
ENE 09: Energy-efficient escalators	2	To promote energy savings and CO_2 reduction through efficient escalators	Verifying that the specifications of the escalators meet the requirements
ENE 26: Assurance of thermal quality of the building	2	To guarantee the thermal quality of the building envelope	Thermographic survey on-site to check for thermal irregularities and quality of insulation
Total max. points	31		

3. Research Methodology

A mix of methods have been applied (shown in Figure 2) in this research, namely desk study, semi-structured interviews, RACI matrix actor mapping, and the development of a custom application to demonstrate the recommendations and validation of these recommendations through semi-structured interviews.

This study starts with a desk study on BIM-based energy performance and green building assessments. A comprehensive qualitative analysis of the requirements of the energy credits stated in the BREEAM-NL new construction and renovation 2014 v2 guideline [29] for building energy retrofits has been performed. These requirements are then mapped to relevant stakeholders, BIM uses, and automation possibilities for each credit. This serves as a starting point for mapping the possibilities of using BIM for automation and also as an input for semi-structured interviews.

This research is practice-oriented, with a focus on the multidisciplinary nature of BREEAM-NL energy evaluations and problems experienced by industry professionals. Therefore, we have embedded this research in a case study organization, and expert interviews have been conducted throughout the process, starting from problem statement initiation, process and actor mapping, and recommendations for energy assessment implementation strategies based on the BIM maturity, toward validation of the proposed strategies in the end. A total of ten semi-structured interviews have been conducted and the duration of each interview ranged from 60 to 75 min to discuss topics listed in Table 3. The interviews have been performed with interviewees' consent and the results have been anonymized.

The semi-structured interviews of actors involved in energy assessments of BREEAM-NL help in four aspects:

- Supplementing and validating the possibilities of using BIM for BREEAM-NL energy credits in building energy retrofits, identified from desk study;
- Understanding the responsible, accountable, consulted, and informed actors in the process and the current implementation status of BIM in BREEAM-NL assessment;
- Proposing customized starting points in response to the current BIM use status for BIM-enabled energy performance evaluation;
- Validating the proposed starting points and the associated applications.

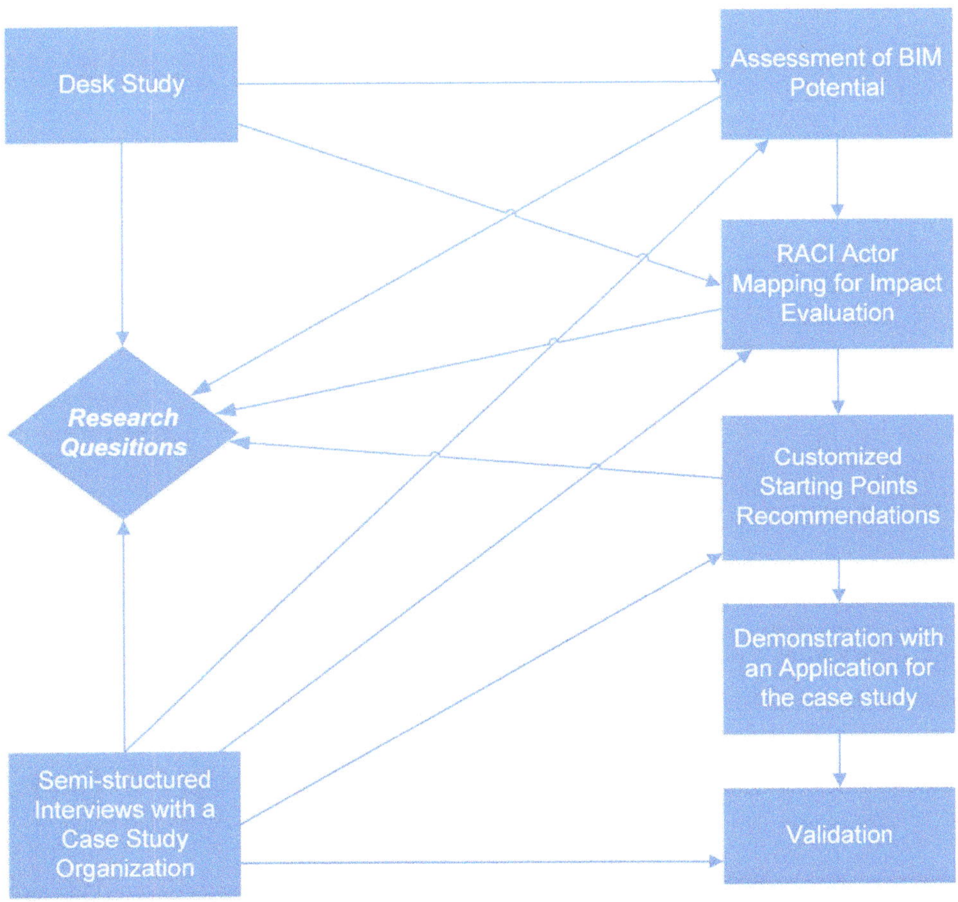

Figure 2. Methodology framework.

Table 3. List of interviewees and topics discussed.

<table>
<tr><th>Actor Code</th><th>Role</th><th>Experience</th><th>Interview Topics</th></tr>
<tr><td>A1</td><td>BREEAM expert and sustainability consultant</td><td>3+ years' experience in BREEAM projects</td><td rowspan="5">(1) Workflow and methodology of energy assessments in BREEAM-NL
(2) Means of information exchange and collaboration with other stakeholders
(3) Awareness and use of BIM for energy assessments
(4) Foreseeable challenges for BIM integration
(5) Time taken for energy assessments
(6) Most critical/time-consuming steps in the process</td></tr>
<tr><td>A2</td><td>BREEAM expert and sustainability consultant</td><td>5 years' experience in BREEAM projects</td></tr>
<tr><td>A3</td><td>BREEAM expert and sustainability consultant</td><td>13 years' experience in BREEAM projects</td></tr>
<tr><td>A4</td><td>Energy specialist</td><td>5 years' experience as Energy consultant</td></tr>
<tr><td>A5</td><td>Energy specialist</td><td>5 years' experience as Energy consultant</td></tr>
<tr><td>A6</td><td>Architect (external actor with a collaboration history with the case study organization)</td><td>20 years' experience in architectural design</td><td rowspan="3">(1) Role and level of involvement in the BREEAM-NL assessment process
(2) Awareness of the BREEAM-NL requirements
(3) Awareness and use of BIM for energy assessments
(4) Foreseeable challenges for BIM integration
(5) Time spent on BREEAM-NL energy assessments
(6) Most critical/ time-consuming steps in the process</td></tr>
<tr><td>A7</td><td>MEP designer</td><td>30 years' experience in MEP design</td></tr>
<tr><td>A8</td><td>MEP designer</td><td>12 years' experience in MEP design</td></tr>
<tr><td>A9</td><td>BIM manager</td><td>9 years' experience with BIM</td><td rowspan="2">(1) Position of the BIM department in the organizational structure
(2) Organizational culture, drivers, and challenges of BIM
(3) Level of involvement in BREEAM-NL projects
(4) BIM maturity level of the organization
(5) BIM tools and processed adopted</td></tr>
<tr><td>A10</td><td>BIM and digitalization lead</td><td>15 years' experience in product development and digitalization</td></tr>
</table>

A responsible, accountable, consulted, and informed (RACI) matrix [30,31] has been made to understand the actors involved and their roles. This serves as the basis for examining the impacts and potential barriers of BIM-enabled workflow on various actors in this process.

To enable the validation of the proposed steps, an application for one example energy credit using REVIT API 2021 has been developed to implement the recommendations. Using the data from the desk study and practical interviews, a custom tool has been developed in C# that can automate the assessment process of the identified example credit.

4. Integration with BIM: Possibilities

After the analysis of the requirements stated in the guideline and interviewing experts, it has been identified that there are two types of data requirements for energy assessments:

- Data about building geometry, characteristics, and specifications of the components used in the building, which can be obtained from BIM models;
- Performance-related data can either be obtained from dynamic BEM models or static calculation methods using the information from BIM models.

For the first category, BIM can be used for intelligent data acquisition and rule checking. Therefore, for the evaluation metrics that concern only this type of information, the assessment can be fully automated if the required information is stored in the BIM models. However, in the case of evaluation metrics that require both building geometry and performance-related data, the integration of BIM data with performance data obtained from software tools approved by the DGBC is required to make energy assessments for BREEAM-NL. Therefore, apart from the availability of data, the interoperability between BIM and BEM tools is detrimental for automation.

For performance-related criteria, two kinds of assessments are possible: dynamic simulations and static calculations. Static methods of calculation make use of pre-defined factors and several assumptions for the assessment of building performance. Building related information for these assessments can be fully extracted from BIM models given an appropriate level of detail (LOD) [32] in the models. These methods are less complex, but fail to capture the impact of environmental and operational dynamics on building performance [33]. Dynamic calculation methods make estimates on building performance considering the dynamic weather conditions, occupant usage, and varying energy demands and, therefore, yield more accurate results. These assessments are carried out in an external energy analysis tool such as Integrated Environmental Solutions Virtual Environment (IESVE) or Green Building Studio (GBS). Table 4 provides an overview of the potential BIM use for automation for the energy credits evaluation.

Table 4. BREEAM-NL energy credits vs. BIM uses.

Credit	BIM Uses	Input Data Source	Automation Possibility
ENE1	Data extraction Performance analysis and prediction Compliance validation	BIM + BEM	Partial
ENE2	Data extraction Documentation Compliance validation	BIM	Full
ENE4	Data extraction Documentation Compliance validation	BIM	Full
ENE5	Data extraction Performance analysis and prediction Compliance validation	BIM + BEM	Partial
ENE6	Documentation Design reviews	BIM	None
ENE7	Data extraction Documentation Compliance validation	BIM	Full
ENE8	Data extraction Documentation Compliance validation	BIM	Full
ENE9	Data extraction Documentation Compliance validation	BIM	Full

5. Integration with BIM: Impacts on Actors

This section starts with a responsible, accountable, consulted, and informed (RACI) matrix developed based on the results from the desk study and the semi-structured interviews from the case study organization (Section 5.1). The RACI matrix is then used to evaluate the potential impacts on actors and implementation in the case study organization (Section 5.2).

5.1. The RACI Matrix for BREEAM-NL Assessment

Table 5 presents the developed RACI matrix for BREEAM-NL assessment in the case study organization, and the BREEAM-NL expert is a special role in this process as described in Section 2.1. It must be noted that these responsibilities relate to design-phase assessment only. The post-construction review is based on as-built information, in which case, BIM can be used to produce documentation.

Table 5. BREEAM-NL energy credits vs. stakeholder responsibilities.

Credit	Input Data for Assessment	Stakeholder				
		BREEAM-NL Expert	**Architect**	**MEP Designer**	**Sustainability Engineer**	**Contractor**
ENE1	EPC Calculation	R				
	Inputs: (a) Geometry and building envelope related information (b) HVAC system details (c) Renewable energy system details		 A I I	 I A A	 C C C	 I I I
	Use of certified calculation software	R				
ENE2	Significant energy consumption zones and placement of energy sub-meters	R	I	A	I	I
	Information on if sub-meters are connected to a building management system	R	I	A	I	I
ENE4	Luminous flux, power, and target illuminance area of lighting fixtures	R	I	A	I	I
	Purpose of lighting fixtures: utility/ decorative	R	C	A	I	I
	Presence of automatic dimming or switching options	R	I	A	I	I
ENE5	Feasibility study and carbon emission calculations	R	I	I	A	I
ENE6	Qualitative assessment of design features	R	A	I	I	I
ENE7	Specification of cold storage equipment	R	I	A	I	I
ENE8	Specifications of elevators	R	I	A	I	I
ENE9	Specifications of escalators and moving walks	R	I	A	I	I
ENE26	On-site thermographic survey and air permeability measurement	R	I	I	I	I

R = responsible for the assessment activity; A = accountable for input information; C = consulted for design optimization; I = informed about design and changes.

In general, the actors marked responsible lead the task of the credit assessment and compliance verification. This is usually the BREEAM-NL expert appointed by the client. They are responsible for the collection of the design data, evaluating the status of compliance in accordance with the BREEAM-NL requirements and creating suitable documentation to do the same. The expert (s) takes the lead role in the assessment process. Starting from performing a quick scan to identify the target BREEAM-NL credits for a project, the expert provides guidance and support throughout the design and construction phases. The responsibility of verifying the compliance with BREEAM-NL requirements and compiling the evidence report to support the same also lies with the expert. In general, the information sources and the assessment tasks for each credit lie with different stakeholders involved in a project.

The actors marked accountable are the authors of the design information such as the mechanical, electrical, and plumbing (MEP) design details used for the EPC calculation in ENE1 or information regarding the energy concept used for the assessment of the credit ENE5. Therefore, the responsibility of providing accurate input parameters for the assessments lies with them. As seen in the brief overview of input information for each credit and the stakeholders responsible for it in Table 5, the primarily accountable stakeholder for most of the energy assessments is the MEP design team. This is because the credit requirements relate mainly to installations or light fixtures, and this information can be found in MEP models or drawings. Whereas, the responsible party for the assessments is the BREEAM-NL expert in all cases. Few criteria in ENE1 and ENE6 relate to geometric information, for which the architect will be accountable.

The actors marked consulted are not primarily involved in the assessment process but rather provide inputs for design optimization in case of non-compliance. For example, to gain a point for ENE 4, all outdoor lighting must be non-decorative. While the lighting design is done by the MEP team, the architect is consulted in the process to ensure that the lighting requirements aligns with the architectural design idea. Sustainability engineers, if appointed for the project, are mainly consulted for credits dealing with energy consumption or carbon reduction.

Additionally, the actors marked informed are not concerned with the design stage assessment itself but the outcome of the assessment influences their scope of work. Such as, the contractor is not a part of the design certification, but all the starting points used for the assessments will be translated into a technical program of requirements for the construction phase. Therefore, they are kept informed of the target credits, status of compliance, and input parameters that cannot be changed during construction.

5.2. RACI-Based BIM Integration Impacts

Based on the RACI model presented in Table 5, besides the expert in the evaluation process, architects and MEP designers are very important, as they are accountable for the inputs used in the energy performance evaluation process for almost all the credits. These are also the actors responsible for the creation of the BIM models and, therefore, have a key role in a BIM-based evaluation process.

However, at this moment, the interview results have indicated that BIM is not being actively used for BREEAM-NL energy assessments in the case study organization (A1–A3). The interviewees expressed that the link between BIM and energy assessments for BREEAM-NL, in particular, is not known to the practitioners in the case study organization (A4). While the other departments of the organizations are making significant progress in BIM adoption, this subject is not considered relevant to the energy team (A2, A9). Knowledge about how many of the credit assessments can be automated and what information must be requested from each stakeholder is not known (A1–A5).

Interestingly, in this case study organization, the players from the BREEAM-NL expert team are not only open but rather enthusiastic about moving to a BIM-enabled partial automatic workflow. They described their current scope of work as being dominated by "administrative tasks" of gathering information from multiple stakeholders, whereas they would rather spend that time on consultancy and advising the clients on how to make their assets more energy efficient (A2, A3).

However, as seen in the RACI matrix, some resistance has also been observed and is also logical from the design team members interviewed, as they will need more time in preparing such information. They have expressed concerns over the increase in modelling time to deliver BIM models that are suitable for automated assessments (A6, A7). Time and budget constraints have been also pointed out as barriers that may prevent its implementation in practice (A2).

Furthermore, it has also been observed from the documentation of BREEAM-NL projects that while most of the projects handled by the BREEAM-NL team do have a BIM execution plan (BEP), this document has not yet included the BREEAM-NL team as a relevant stakeholder, nor has it described their requirements, deliverables, and responsibilities. This is because BIM for sustainability assessments is not yet a commonly adopted strategy in practice, and therefore, there is not sufficient guidance available to help execute the same. The interviews have also revealed that the BIM models handed over by the design teams are often not suitable for energy analysis. Due to poor interoperability between BIM and BEM tools, the model has to be either cleaned or re-modelled before running an energy simulation (A2, A5). This process is highly time consuming. Since dynamic simulations are not mandatory, static methods are often used. However, even in this case, all the information required for assessments is not present in the BIM models and is often requested through email or other digital communication platforms (A4).

Therefore, to implement a BIM-enabled workflow for energy performance evaluation, it is crucial to obtain these actors' support and link the assessment team with them at the beginning of the process with established guidelines on how to work together. These findings also show that support from the client, contractual changes, and comprehensive policy documents guiding the integration of BIM and sustainability assessments will be needed for a successful implementation.

There has been also the question of to what extent the switch to a fully BIM-based workflow is justified. Interviewees have pointed out that for critical or time-consuming credits, the use of BIM has obvious advantages such as significant improvements in building performance and efficiency of the design process, which would be welcomed by all the project stakeholders. However, some interviewees have held the opinion that not all the credits that can be automated using BIM ***need*** to be automated (A3, A7, A8). Credits such as ENE8 or ENE9 relate to the use of energy-efficient transport equipment. It is theoretically possible to include this information within BIM models and create rules that can verify the compliance of a given model against them. However, the specific requirements stated in the assessment guideline are numerous. The amount of effort that goes into the creation of the automation infrastructure and information-rich models does not outweigh the benefits in this case. It is also important to consider the current level of BIM adoption in a team before proposing strategies for digitalization and automation of assessments. Progress to higher BIM maturity can be met with resistance if it is not approached in small incremental steps (A9, A10). Therefore, from a practical perspective, it is lucrative to identify the most critical topics of energy evaluations for the switch to a BIM-based approach. As a result, customized recommendations are given to the case study organization based on their current practice status of BIM and BREEAM-NL assessment for energy performance assessment.

6. Customized Recommendations

6.1. Energy Performance Evaluation with BIM—Starting Points

The extent to which BIM-based energy assessments are possible greatly depends upon the BIM maturity stage [23,34] the project team operates at. The benefits offered by BIM and integration with sustainability assessments increase with each stage [5,26]. As can be seen from Table 5, most of the energy credits involve multiple stakeholders, and therefore, real-time information exchange between these disciplinary models is required for the automation of assessments for these credits. This is only possible at a higher BIM stage. Whereas, for credits that have a single information source, it is possible to automate the assessments at lower BIM stages provided that the disciplinary models are information rich. Therefore, when approaching the automation of performance assessments for energy efficiency and green certifications, it is important to consider the stakeholder in question as well as the BIM maturity of the project. Furthermore, apart from the technological infrastructure, it is also important to consider the expected change in working methodologies due to automation and the guidelines for ensuring a successful implementation, as mentioned in Section 5.2.

The used case study organization acts as a BREEAM-NL consultant in multidisciplinary architecture, engineering, and construction (AEC) projects. It has been identified that this team is currently in the nascent stages of BIM adoption (A9, A10). The organization also specializes in mechanical, electrical, and plumbing (MEP) design, and the MEP team is also in the same organization as most of the projects dealt with by the BREEAM-NL team. Based on this information and the above-mentioned findings, the following recommendations are provided for the case study organization for the transition to BIM-enabled energy performance assessments:

- Identify target credits for automation: BIM offers varying levels of benefits depending on the requirements of the evaluation criteria. Therefore, before investing time and effort in developing software infrastructure for automating the assessment process, it is recommended to identify the credits that have the greatest impacts on design performance or are the most time-consuming ones. These credits will be taken further for automation.
- Identify the accountable stakeholders for information related to these credits. Making agreements on MEP-related information is relatively easier for this case, as it is an internal department. For credits that involve information related to other disciplines, these information requirements must be specified at the beginning of a project in the BIM execution plan.
- Aligning the input requirements of the automation tools with the existing data exchange formats and workflows wherever possible would help lead to a smoother adoption. Additionally, the organization should already start with developing guidelines on collaboration among teams.

6.2. Demonstration with One Selected Example Credit

This section follows the customized recommendations to the case study organization and the desired credits to be automated based on the desk study and interviews.

According to the current BIM adoption level of the organization and the organizational arrangement for the BREEAM-NL assessment process presented in the RACI matrix, ENE1, ENE4, and ENE5 have been identified as the target credits to use the BIM-based automation process. Of these three credits, ENE4 is chosen for the demonstration, as this one only requires MEP-related information, and in this organization, MEP designers are in the same team as the BREEAM-NL assessment members. This credit aims to promote the use of energy-efficient lighting fixtures in the exterior areas. The BREEAM-NL assessment guideline specifies three criteria for this credit:

- The luminous flux from the luminaires must be more than 65 lumen/watt and the maximum allowable specific power per lux illumination is 0.1 W (lux/m^2).
- Decorative or mood lighting with a non-renewable energy source is not allowed.
- Light fixtures must be equipped with automatic switching or dimming features.

An Autodesk REVIT-based plugin has been developed using REVIT API 2021 to automate the assessment for this credit using BIM data. Since no existing BIM tool has the BREEAM-NL requirements integrated in it, a platform that allows for the development of a custom plugin was needed, which is offered by REVIT API. This is also the software commonly used by both BREEAM-NL and MEP design teams (A7–A10) and, therefore, is considered the suitable choice for the practical implementation. The algorithm has made use of two built-in REVIT parameters namely luminous flux and wattage. In addition to this, three custom parameters have been defined for the lighting fixture family, namely exterior lighting, sensor connection, and mood lighting, as shown in Table 6. The lighting fixtures belonging to a zone have been grouped. Subsequently, spaces have been created matching the lighting group names. A custom parameter for spaces with the name "target illuminance" has been defined.

Table 6. Parameters used in REVIT ENE4 plugin.

Parameter Name	Parameter Type	Category	Value Type
Luminous flux	Built in	Lighting fixtures	Integer
Wattage	Built in	Lighting fixtures	Integer
Exterior lighting	Custom defined	Lighting fixtures	Y/N
Mood lighting	Custom defined	Lighting fixtures	Y/N
Sensor connection	Custom defined	Lighting fixtures	Y/N
Target illuminance (In Lux)	Custom defined	Spaces	Integer

The sequence of steps and decision gates used in the plugin is shown in Figure 3.

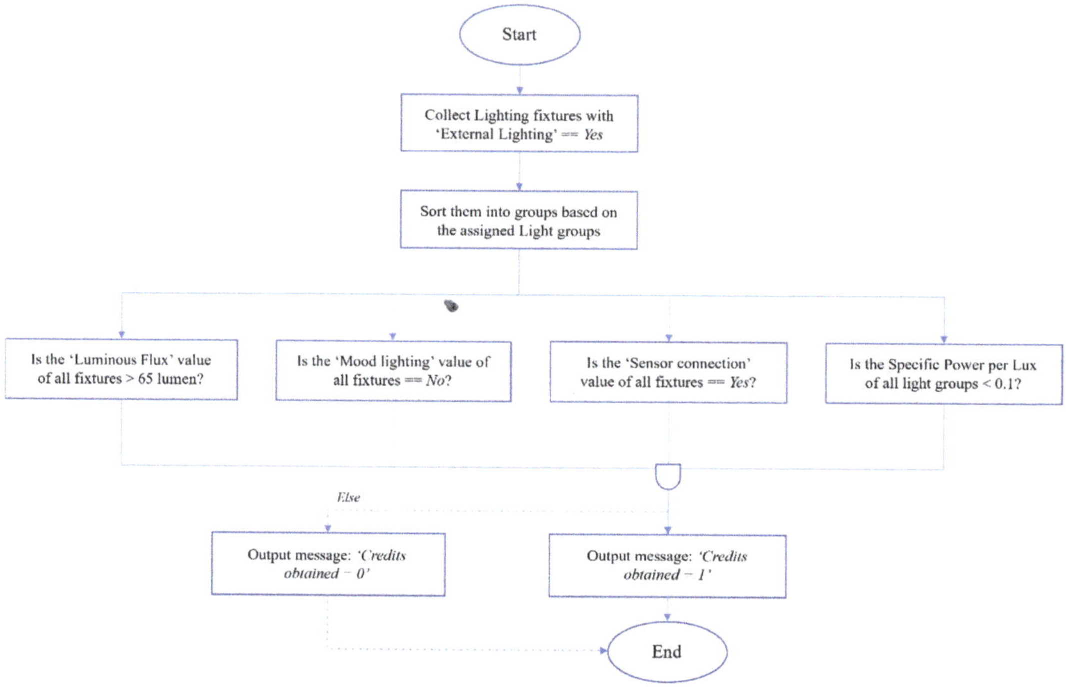

Figure 3. ENE4 plugin flowchart.

The plugin first filters through all the lighting fixtures present in the file using the parameter "exterior lighting". Then, the values given for luminous flux, mood lighting, and sensor connection are collected. Following this, specific power per lux calculation is performed for each lighting group using the properties of lighting fixtures and the area and target illuminance values of the associated space. Finally, if all the three BREEAM-NL criteria are met, the plugin provides a score of 1 point along with a report generated in CSV format.

The plugin has been tested on a sample REVIT model for the validity of the results (see Table 7 and Figure 4). The underlined parts indicate the not qualified criteria in the assessment. The algorithm displays a brief overview of the results in the task dialogue, and a detailed report is exported to a predefined location. This report can be used for documentation for evidence submission.

Table 7. Report generated by the plugin.

Group	Fixture Name	Mood Lighting	Sensor Connection	Luminous Flux	Watt	Light Number	Total Power	Area	Target Illuminance	Power Per Lux
Outdoor Zone 1	400-watt Halogen 2	Yes [1]	No [2]	4500	250	2	500			
Outdoor Zone 1								38	25	0.5263158
Outdoor Zone 2	Street Light-Sgl	No	Yes	5005	278	7	1946			
Outdoor Zone 2	400-watt Halogen 2	Yes [1]	No [2]	4500	250	1	250			
Outdoor Zone 2								64	13	2.639.423 [3]

[1] Outdoor lighting for decorative purposes leads to non-compliance and a score of 0. [2] Lack of automatic dimming or switching option leads to energy wastage and therefore is non-compliant with the credit requirements. Leads to a score of 0. [3] The maximum allowable specific power per lux illumination is 0.1 W (lux/m^2), and any value above this leads to a score of 0.

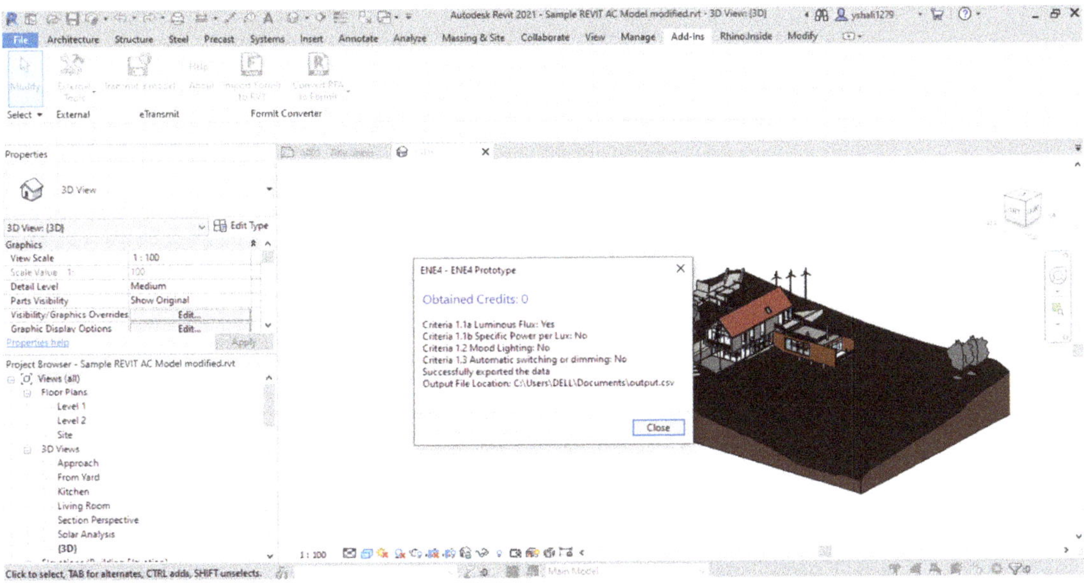

Figure 4. Screenshot of plugin results.

The results have been compared with the ones obtained through manual assessment methods. The reason for not using a real case study is that none of the existing models contained the information required for assessments. From the interviews, it has been found that it takes about 4 h for the BREEAM-NL team each time the assessment has to be made. Provided that the required information is in the BIM Models, this time can be reduced to just ten seconds, and the assessments can be made at any point in the design process.

7. Conclusions and Discussion

To accelerate building energy retrofits in the Netherlands, this study explores the potential of using BIM for automating energy performance evaluation for BREEAM-NL certification. The study used a case study organization to better understand the complexities of the assessment process and challenges faced in the industry concerning the application of BIM for building energy retrofitting evaluation.

This study combines literature review and practice research using semi-structured interviews to understand first the current workflow for BREEAM-NL energy assessment for the building retrofit process, and then, map this process with the BIM uses to identify possibilities for automation. A RACI matrix is developed to map different actors involved in the process, and potential impacts on actors involved are presented if automating energy credits evaluation will be implemented.

Based on the context in the case study organization and the BIM maturity of the sustainability team in the case study organization, customized recommendations are proposed for the implementation of BIM-based energy evaluations. It has been identified that in the context of the case study organization, which specializes in sustainability and MEP design consultancy, targeting ENE1, ENE4, and ENE5 credits for BIM-based automation is the most lucrative. A plugin is developed to demonstrate the automation process for the one identified example energy credit, and experts are used to validate the approach. The demonstration has revealed that the BIM-based automation can save a significant amount of time in the assessment process and can also improve the designs by providing an option of performance evaluation at any point in the design process.

The novelty of this study lies in the following aspects:

- Providing an overview for BIM-enabled automation potential for different energy credits in the BREEAM-NL assessment based on literature and practical interviews;
- Linking the changes in workflow due to BIM integration with the actors involved to identify potential impacts for each actor in different aspects;
- Offering a customized strategy to facilitate the implementation of BIM-based evaluations based on the contextual findings from the case study organization.

In countries such as the Netherlands, where a major chunk of the existing building stock is over 50 years old, renovation and retrofitting projects are becoming very essential to meet the energy requirements of the future. The benefits of using BIM on the sustainability performance evaluation of a building have been extensively discussed in the literature. BIM is known to improve the design process by allowing for a simultaneous evaluation of its performance and alternate solutions. However, the status of BIM-based sustainability and energy efficiency assessments as noticed in this case study team is lagging. BIM-based energy efficiency or sustainability evaluations for green certifications are not recognized as a common BIM use in the industry. This is reflected in the BIM execution plans, where neither the relevant BIM uses are identified nor the certification experts listed as relevant stakeholders. Based on the literature review, for the credits in the energy category of BREEAM-NL, intelligent data gathering through quantity take-offs, scheduling, performance analyses, and code compliance verification are the possible BIM uses. However, in practice, BIM models are being merely used as design references.

The findings of this study show that one of the primary reasons for this gap is the multidisciplinary nature of BREEAM-NL certification projects. Digitalization requires a considerable amount of preparatory efforts as it is, but even more so in the context of the transformation of intra-organizational processes. Development of automation infrastructure or digitalization strategies must bear this in mind. At an inter-organizational level, the transition to a BIM-based assessment process can be approached by identifying the credits that are most critical and relevant for the disciplinary specialization of the organization. The data requirements for automated assessments must be communicated to project stakeholders at the beginning of the design process. The automation process depends largely on the BIM maturity level of an organization, and therefore, context-specific strategies need to be taken. The process followed in this study can be used for other organizations to analyze their BIM possibilities and design digital transition strategies accordingly.

There are several limitations of this research that requires further exploration. First, the scope of the study is limited to design-phase assessments for BREEAM-NL certification. Further study can extend to other phases and see how to link lifecycle asset management to BIM use in energy performance evaluations. Second, this study builds on the hypothesis of the availability of a BIM model of the existing asset. The focus of the study lies in the applicability of BIM data for further analysis of building performance, and therefore, the process of data acquisition is not focused upon. For the existing building stock, not a lot of BIM models exist. A potential research direction could be integrating scan to BIM technologies and using voxelized geometries to automate building model processing process and generating suitable formats for BEM tools. Third, real projects are probably more complicated, which needs further research and validation to evaluate the efficiency gains with such a BIM-enabled automation process. Other types of categories in the BREEAM-NL assessment could be automated as well, and the overall efficiency gains could be calculated. Fourth, a holistic evaluation framework for stakeholders' involved could be developed to make better business decisions based on the data gathered such as efficiency gains and cost–benefit analysis. Fifth, as indicated in Section 5.2, the starting points for an organization to transit toward a BIM-based assessment process depends not only on technology readiness but also on the organizational business process and people involved. Therefore, a BIM maturity matrix that combines process, technology, and people aspects can be developed for different organizations to evaluate their BIM status and further design their digitalization strategies for not only BREEAM-NL but also other certification systems.

Author Contributions: Conceptualization, all authors; methodology, V.S.; formal analysis, V.S. and T.W.; writing—original draft preparation, V.S. and T.W.; writing—review and editing, T.W., Y.L. and H.W.; supervision, T.W., Y.L., H.W., L.M. and G.v.G. All authors have read and agreed to the published version of the manuscript.

Funding: This research receives APC funding from TU Delft.

Institutional Review Board Statement: The study was conducted according to the guidelines of the Declaration of Helsinki. Ethical review and approval were waived for this study. Based on the guidelines of TU Delft research by students, the researchers concluded that the research isn't considered as extensive risk.

Informed Consent Statement: Informed consent was obtained from all the subjects involved in this study.

Data Availability Statement: The data presented in this study come from the DGBC and Deerns.

Acknowledgments: We acknowledge the support from Deerns and all the comments from the reviewers.

Conflicts of Interest: The authors declare no conflict of interest.

References

1. Paris Agreement 2015. Available online: https://unfccc.int/sites/default/files/english_paris_agreement.pdf (accessed on 2 December 2021).
2. Rijkswaterstaat National Climate Agreement 2019. Available online: https://www.government.nl/documents/reports/2019/06/28/climate-agreement (accessed on 2 December 2021).
3. European Commission 2017. Available online: https://ec.europa.eu/clima/eu-action/climate-strategies-targets/2030-climate-energy-framework_en (accessed on 2 December 2021).
4. Rijksoverheid Stimuleert Duurzame Energie 2020. Available online: https://www.rijksoverheid.nl/onderwerpen/duurzame-energie/rijksoverheid-stimuleert-energiebesparing (accessed on 2 December 2021).
5. Carvalho, J.P.; Bragança, L.; Mateus, R. A Systematic Review of the Role of BIM in Building Sustainability Assessment Methods. *Appl. Sci.* **2020**, *10*, 4444. [CrossRef]
6. Kuistermans, C. *Sustainability—Does It Influence Investors' Decision? An Exploration to BREEAM, as Part of a Multi-Criteria Decision Analysis*; Technische Universiteit Eindhoven: Eindhoven, The Netherlands, 2012.
7. Van Eldik, M.A.; Vahdatikhaki, F.; dos Santos, J.M.O.; Visser, M.; Doree, A. BIM-Based Environmental Impact Assessment for Infrastructure Design Projects. *Autom. Constr.* **2020**, *120*, 103379. [CrossRef]
8. Basbagill, J.; Flager, F.; Lepech, M.; Fischer, M. Application of Life-Cycle Assessment to Early Stage Building Design for Reduced Embodied Environmental Impacts. *Build. Environ.* **2013**, *60*, 81–92. [CrossRef]
9. Bueno, C.; Pereira, L.M.; Fabricio, M.M. Life Cycle Assessment and Environmental-Based Choices at the Early Design Stages: An Application Using Building Information Modelling. *Archit. Eng. Des. Manag.* **2018**, *14*, 332–346. [CrossRef]
10. Bueno, C.; Fabricio, M.M. Comparative Analysis between a Complete LCA Study and Results from a BIM-LCA Plug-In. *Autom. Constr.* **2018**, *90*, 188–200. [CrossRef]
11. Krygiel, E.; Nies, B.; McDowell, S. Green BIM: Successful Sustainable Design with Building Information Modeling. In *Green BIM: Successful Sustainable Design with Building Information Modeling*; Wiley Publishing, Inc.: Indianapolis, Indiana, 2008.
12. Lu, Y.; Wu, Z.; Chang, R.; Li, Y. Building Information Modeling (BIM) for Green Buildings: A Critical Review and Future Directions. *Autom. Constr.* **2017**, *83*, 134–148. [CrossRef]
13. Motawa, I.; Carter, K. Sustainable BIM-Based Evaluation of Buildings. *Procedia Soc. Behav. Sci.* **2013**, *74*, 419–428. [CrossRef]
14. Alsehrawy, A.; Tong, M.; Callaghan, N.; Amoudi, O. A Critical Review of the Challenges and Barriers to an Effective Use of BIM in Green Building Assessment. In Proceedings of the Construction Digitalisation for Sustainable Development: Transformation through Innovation, Hanoi, Vietnam, 24–25 November 2020.
15. Ayman, R.; Alwan, Z.; McIntyre, L. *Factors Motivating the Adoption of BIM-Based Sustainability Analysis*; LSI Publishing: Dublin, Ireland, 2018.
16. Ade, R.; Rehm, M. The Unwritten History of Green Building Rating Tools: A Personal View from Some of the 'Founding Fathers'. *Build. Res. Inf.* **2020**, *48*, 1–17. [CrossRef]
17. Sánchez Cordero, A.; Gómez Melgar, S.; Andújar Márquez, J.M. Green Building Rating Systems and the New Framework Level(s): A Critical Review of Sustainability Certification within Europe. *Energies* **2019**, *13*, 66. [CrossRef]
18. Wu, W.; Issa, R.R.A. BIM Facilitated Web Service for LEED Automation. In *Proceedings of the Computing in Civil Engineering (2011), Miami, FL, USA, 19–22 June 2011*; American Society of Civil Engineers: Reston, VA, USA; pp. 673–681.
19. Jiang, S.; Wang, N.; Wu, J. Combining BIM and Ontology to Facilitate Intelligent Green Building Evaluation. *J. Comput. Civ. Eng.* **2018**, *32*, 04018039. [CrossRef]

20. Dutch Green Building Council BREEAM-NL 2010 Keurmerk voor duurzame vastgoedobjecten: Beoordelingsrichtlijn Nieuwbouw, Rotterdam, the Netherlands. Available online: https://cupdf.com/document/20100929brl2010-v20breeam-nl-nl-55ab4ed862207.html (accessed on 2 December 2021).
21. Long-Term Renovation Strategy: En Route to a Low-CO2 Built Environment 2020. Available online: https://ec.europa.eu/energy/sites/ener/files/documents/nl_2020_ltrs_en.pdf (accessed on 2 December 2021).
22. Saheb, Y.; Bódis, K.; Szabó, S.; Ossenbrink, H.; Panev, S. Energy Renovation: The Trump Card for the New Start for Europe. In *European Commission-Joint Research Centre-Institute for Energy and Transport, JRC Science and Policy Reports*; Publications Office of the European Union: Luxembourg, 2015.
23. Succar, B.; Sher, W.; Williams, A. Measuring BIM Performance: Five Metrics. *Archit. Eng. Des. Manag.* **2012**, *8*, 120–142. [CrossRef]
24. Borrmann, A.; König, M.; Koch, C.; Beetz, J. (Eds.) *Building Information Modeling: Technology Foundations and Industry Practice*; Springer International Publishing: Cham, Switzerland, 2018; ISBN 978-3-319-92861-6.
25. Pereira, V.; Santos, J.; Leite, F.; Escórcio, P. Using BIM to Improve Building Energy Efficiency—A Scientometric and Systematic Review. *Energy Build.* **2021**, *250*, 111292. [CrossRef]
26. Carvalho, J.; Almeida, M.; Bragança, L.; Mateus, R. BIM-Based Energy Analysis and Sustainability Assessment—Application to Portuguese Buildings. *Buildings* **2021**, *11*, 246. [CrossRef]
27. Zanni, M.-A.; Soetanto, R.; Ruikar, K. Defining the Sustainable Building Design Process: Methods for BIM Execution Planning in the UK. *Int. J. Energy Sect. Manag.* **2014**, *8*, 562–587. [CrossRef]
28. Ansah, M.K.; Chen, X.; Yang, H.; Lu, L.; Lam, P.T.I. A Review and Outlook for Integrated BIM Application in Green Building Assessment. *Sustain. Cities Soc.* **2019**, *48*, 101576. [CrossRef]
29. Dutch Green Building Council BREEAM-NL Niewubouw en Renovatie- Keurmerk voor Duurzame vastgoedobjecten: Beoordelingsrichtlijn 2014 versie 2. Available online: https://www.google.com.hk/url?sa=t&rct=j&q=&esrc=s&source=web&cd=&cad=rja&uact=8&ved=2ahUKEwiH-6O44tD0AhWOl2oFHcRmBnoQFnoECAYQAQ&url=https%3A%2F%2Frichtlijn.breeam.nl%2Fupload%2Ffiles%2FNieuwbouw%2520en%2520Renovatie%2FVersie%25202.0%2FBRL%2520BREEAM-NL%2520nieuwbouw%25202014_v2.pdf&usg=AOvVaw1ztyYmdlliMkBqJZrxlFFT (accessed on 2 December 2021).
30. Project Management Institute (Ed.) *A Guide to the Project Management Body of Knowledge (PMBOK Guide)*, 5th ed.; Project Management Institute, Inc.: Newtown Square, PA, USA, 2013; ISBN 978-1-935589-67-9.
31. Jacka, J.M.; Keller, P.J. *Business Process Mapping: Improving Customer Satisfaction*, 2nd ed.; Wiley: Hoboken, NJ, USA, 2009; ISBN 978-0-470-44458-0.
32. Tang, L.; Li, L.; Ying, S.; Lei, Y. A Full Level-of-Detail Specification for 3D Building Models Combining Indoor and Outdoor Scenes. *ISPRS Int. J. Geo-Inf.* **2018**, *7*, 419. [CrossRef]
33. Jradi, M. Dynamic Energy Modelling as an Alternative Approach for Reducing Performance Gaps in Retrofitted Schools in Denmark. *Appl. Sci.* **2020**, *10*, 7862. [CrossRef]
34. Succar, B. Building Information Modelling Framework: A Research and Delivery Foundation for Industry Stakeholders. *Autom. Constr.* **2009**, *18*, 357–375. [CrossRef]

energies

MDPI

Article

Heat Pumps, Wood Biomass and Fossil Fuel Solutions in the Renovation of Buildings: A Techno-Economic Analysis Applied to Piedmont Region (NW Italy)

Edoardo Ruffino, Bruno Piga, Alessandro Casasso * and Rajandrea Sethi

Politecnico di Torino—Department of Environment, Land and Infrastructure Engineering (DIATI), Corso Duca Degli Abruzzi 24, 10129 Torino, Italy; ruffino.edoardo@gmail.com (E.R.); bruno.piga89@gmail.com (B.P.); rajandrea.sethi@polito.it (R.S.)

* Correspondence: alessandro.casasso@polito.it; Tel.: +39-3204213886

Abstract: The levelized cost of heat (LCOH) and the technical feasibility in the specific context of building construction or renovation are the major drivers of users' choices for space heating and cooling solutions. In this work, the LCOH was assessed for the most diffused heating technologies in Piedmont (NW Italy): that is, fossil fuels (methane, heating oil and liquefied petroleum gas—LPG), wood biomass (wood logs and pellet) and heat pumps (air-source and ground-source), both in heating-only and in a heating and cooling configuration. A sensitivity analysis of the main LCOH drivers was performed to assess whether and how each technology is vulnerable to energy price and upfront cost changes. The results show that heat pumps are competitive against gas boilers, but they are heavily dependent on refurbishment incentives and penalized by the high electricity prices in Italy; on the other hand, wood biomasses are competitive even in the absence of incentives. The analysis confirmed that LPG and heating oil are no more competitive with renewable heating. Acting on the taxation of natural gas and electricity is key to making heat pumps the most economically convenient solution to cover the heating and cooling needs of buildings.

Keywords: LCOH; life-cycle cost; heating; cooling; heat pump; fossil fuel; biomass; greenhouse gas

Citation: Ruffino, E.; Piga, B.; Casasso, A.; Sethi, R. Heat Pumps, Wood Biomass and Fossil Fuel Solutions in the Renovation of Buildings: A Techno-Economic Analysis Applied to Piedmont Region (NW Italy). *Energies* **2022**, *15*, 2375. https://doi.org/10.3390/en15072375

Academic Editors: Cynthia Hou and Joseph H. K. Lai

Received: 7 February 2022
Accepted: 21 March 2022
Published: 24 March 2022

1. Introduction

In July 2021, the European Union adopted a challenging plan for cutting emissions, called "Fit for 55", as part of the European Green Deal [1]. "Fit for 55" aims to reduce greenhouse gas (GHG) emissions by at least 55% by 2030, compared to 1990 levels, as a step towards becoming the first climate-neutral continent by 2050. The foreseen measures are the consolidation of already existing systems, such as the EU Emissions Trading System [2], brand new mechanisms, such as the Carbon Border Adjustment Mechanism [3], and energy efficiency policies, combined with a significant increase in production from renewable energy sources (RES). The new Renewable Energy Directive (RED 2) increased the target of RES share in 2030 from 32% (as set in 2018 by the RED) to 40% [4]. Moreover, the Energy Efficiency Directive set a 39% reduction in primary energy consumption compared to 1990 through increasing efficiency. Therefore, RES are a cornerstone of European climate policies, and the EU population is aware of their importance. However, citizens see energy prices and their stability as a source of concern and, despite the fact that RES often outperform fossil fuels in the production of electricity in terms of the levelized cost of energy (LCOE) [5], the high initial investment required hinders the diffusion of renewables. The LCOE and its heat counterpart, the LCOH (levelized cost of heat), are widely used indicators of energy costs based on operational expenditure and initial investment, distributing the cost over the lifetime energy production with a discounting method [6]. The initial investment required to produce electricity from RES has decreased at a much faster pace than that for fossil fuels-based power production. For example, the LCOE of utility-scale solar photovoltaics decreased by 89% from 2009 to 2019, compared to a 33% reduction observed for gas combined cycle turbines in the same period [7,8]. Such a significant cost reduction is related to the "learning curve" of renewables, i.e., an effect of increasing installations [7,9], and this effect is therefore likely to continue in the future.

While the economic viability of electrical RES is acknowledged, this does not always apply to thermal RES for space heating and domestic hot water (DHW) production [10]. The studies available in the literature are fewer compared to those on power production. Doračić et al. (2018, 2020) presented the concept of the levelized cost of excess heat to assess the economic viability of recovering waste heat to supply a district heating network and tested it in the Croatian city of Ozalj [11,12]. Li et al. (2019) used the LCOH to find the optimal combination of CHP, biomass, and bio-oil boilers [13]. Vivian et al. (2018) evaluated the LCOH of booster heat pumps to be installed as substations in low-temperature district heating networks (15 °C to 45 °C) [14]. The economic analysis highlighted that this solution is already competitive with natural gas boilers if the low-temperature heat source is sufficiently cheap. The results of cost comparisons between technologies are often country-dependent, e.g., in the United Kingdom (UK), the LCOH of a gas boiler is lower than both an air-source (ASHP) and ground-source heat pump (GSHP) [15–17], whereas the opposite occurs in France [18], although the countries have a similar climate [19]. As pointed out by the European Heat Pump Association [20], the ratio between the unit cost of electricity and gas is critical for heat pump diffusion: the lower this ratio is, the easier it is for heat pumps to penetrate the market. Indeed, based on Eurostat data as of 2020 [21], this ratio is equal to 2.63 in France and 4.63 in the UK.

Although they are not always the most economically convenient solution, heat pumps are expected to play a relevant role in the energy transition. For instance, the Italian Integrated National Energy and Climate Plan (INECP) aims to reach 5.7 Mtoe of heat pump heat production by 2030, i.e., 12.8% of the heating demand compared to the current 4.7% (2.65 Mtoe) [22]. This choice is due to several reasons, including: (i) the wide range of applications of heat pumps in existing buildings, (ii) the absence of any pollutant emissions on site, which makes them suitable for urban areas, and (iii) their possible role as a flexible regulator of the electrical grid [23–25]. The economic convenience of heat pumps, however, is key to achieving the expected growth of this technology. As well as the above-described cost ratio between electricity and gas, the other key factor is the presence of incentives. As pointed out in several studies, the most effective way to boost the energy efficiency of buildings is represented by incentives and standards and, in particular, by performance-based incentives [26–28], whereas informative campaigns are hardly effective. Fewer studies are available on the effect of such policies on the adoption of renewable heating technologies; however, a similar effect of incentives is expected since, in both cases, they increase the economic viability of the targeted interventions.

The studies presented above provide insights into the economic viability of heating and cooling technologies. However, a recent and comprehensive comparison is missing, especially for detached houses and blocks of flats, since the few articles addressing LCOH generally deal with district heating or, when dealing with individual plants, they consider a few heating and/or cooling technologies (e.g., the aforementioned studies on heat pumps vs. gas boilers). This study aims to fill this gap and focuses on the technical feasibility and the economic viability of fossil fuels (methane, heating oil and LPG), wood biomass (logs and pellet) and heat pumps (aerothermal and geothermal) for the heating, cooling and DHW supply in the context of the refurbishment of existing buildings. In the case of heat pump, a photovoltaic (PV) system was included in the analysis and sized to meet the electrical demand of the heat pump. Existing buildings were chosen because, due to the slow renewal rate of building stock [29], they are the key to the decarbonization of the housing sector. The renovation was assumed to be realized on a single detached house and an apartment block, which are the most widespread types of buildings in Northern Italy and, especially, in Piedmont [30–32]. Two settlements representing the typical climates of the Piedmont Region, Turin and Oulx (temperate and cold continental climate, respectively), are the selected spots where model buildings are located. Our analysis included the analysis of the impact of incentives on the LCOH values of each technology, as well as the effect of considering or not the space cooling demand in the calculation of the LCOH. In addition, a sensitivity analysis was performed to evaluate the impact of variable parameters, i.e., the unit costs of energy, the installation costs of the main components and the interest rate used for discounting. This sensitivity analysis permits extending the validity of results for territories other than Piedmont (NW Italy) but with a comparable climate.

The paper is structured as follows. Section 2 describes the methodology adopted, i.e., the selection of representative case studies, the assessment of the energy demand of the buildings in different climatic conditions, the sizing of the heating and cooling systems with different techniques, and the estimation of the installation and operational costs. Section 3 presents the results, focusing on life-cycle costs and their sensitivity to possible future changes in the most influential parameters. Section 4 reports a discussion with a comparison of the results with other studies and with some insights on the diffusion of renewable heating technologies. Conclusions and policy implications are reported in Section 5.

2. Methods

The economic analysis conducted in this article is based on the thermal needs simulated for two benchmark buildings, namely a single detached house and an apartment block of 10 flats. The energy simulation of benchmark buildings and the estimation of the related thermal needs (heating, cooling and DHW) are discussed in Section 2.1. Thermal needs were used as the input for the plant sizing and the assessment of fuel demand (Section 2.2), which are, respectively, the input for estimating the initial investment (Section 2.3) and operating costs (Section 2.4). The levelized cost of heat (LCOH) was chosen as the indicator to identify the most economically convenient technologies, and the estimation method is described in Section 2.5.

The boundaries of all the economic analyses are represented by the heating, DHW and cooling systems (if any), plus the possible replacement of heating/cooling terminals. The interventions on the building envelope were not considered because this analysis aims to assess, case-by-case, the economic viability of heating and cooling technologies with the aim of supporting policies for RES-based technologies.

2.1. Benchmark Buildings and Thermal Needs Assessment

2.1.1. Choice of Representative Buildings and Locations

The techno-economic analysis was based on two representative buildings with constructive characteristics, chosen according to a review of studies on the building stock in Italy and, especially, Piedmont (NW Italy). Two EU-funded projects, TABULA and EPISCOPE [30,31], provide detailed datasets on residential buildings of different ages [32]. According to these references, 35% of buildings in Piedmont were built in the 1960s and the 1970s, and only 18% later on. For this reason, a single detached house and a small block of flats (10 apartments) with building envelope characteristics typical of the 1970s were selected as benchmarks for our study. According to TABULA and EPISCOPE [30,31], these benchmark buildings are the most representative types in Piedmont from the point of view of both structural and envelope thermal characteristics. The assigned thermal transmittance values (hereby, U-values, $Wm^{-2}K^{-1}$) of different elements of the building envelope are reported in Table 1 and compared with current legislative requirements [33], which are aimed at reducing the heating loads and, thus, at saving energy.

Table 1. Thermal transmittance (U-value, $Wm^{-2}K^{-1}$) of the benchmark buildings of this study, based on the TABULA/EPISCOPE databases [30,31], along with the current legislative requirements [33] considered for the simulation of fully refurbished buildings.

Component	Original U-Value	Climatic Zone	Required U-Value
Upper roof	1.65	E	0.20
		F	0.19
Lower floor	1.30	E	0.25
		F	0.23
Perimetral walls	1.26	E	0.23
		F	0.22
Windows	2.8	E	1.30
		F	1.00

Two types of refurbishments were hypothesized for these two buildings: a partial renovation, i.e., replacing only windows and the HVAC system, and a complete renovation of the building envelope, i.e., including an external insulation coating of external walls. Both these interventions are eligible for tax deductions, but they must comply with the legislative requirements on the transmittance of building envelope elements (U-values reported in Table 1). The partial renovation is performed, for example, on buildings where interventions on external walls are not possible due to architectural or technical constraints. On the other hand, a complete renovation makes old buildings almost as well-insulated as brand-new ones due to the very demanding Italian legislative requirements, and, as shown later, this results in a significant difference in thermal needs. Legislative requirements in Table 1, which are given by the regulation on the eligibility of building refurbishment interventions for tax deductions [33], show slightly different U-values depending on the climatic zones set by the Italian DPR 412/93 [34]. The whole territory of Piedmont falls into the two coldest zones if Italy, namely "E" and "F". For this reason, two representative climate data sets were chosen: the capital Turin (temperate continental climate) and the mountain village of Oulx (cold continental climate); based on the Italian norms, the heating degree-days (HDD) are, respectively, 2617 and 4100, calculated with a reference indoor temperature of 20 °C.

Figures 1 and 2 report the plan view of the benchmark buildings analyzed. They are modelled based on the geometrical construction characteristics of buildings suggested by the projects TABULA and EPISCOPE [30]. In more detail, the reference states that an average single detached house from the 1970s has a net floor area of 156 m^2 and a typical small apartment block of 10 housing units has an overall floor area of 934 m^2 (i.e., 187 m^2 per floor). The shape factors (surface over volume ratio) are S/V = 0.73 and S/V = 0.54, respectively. The higher the S/V value, the higher the impact of the transmittance of the building external envelope. For this reason, the thermal needs per m^2 of the apartment block (which has a more compact shape) are expected to be lower than those of the single detached house, being insulation equal. Each apartment is considered a singular thermal zone; therefore, the layout of the internal walls (Figure 2) does not affect the computation of thermal needs.

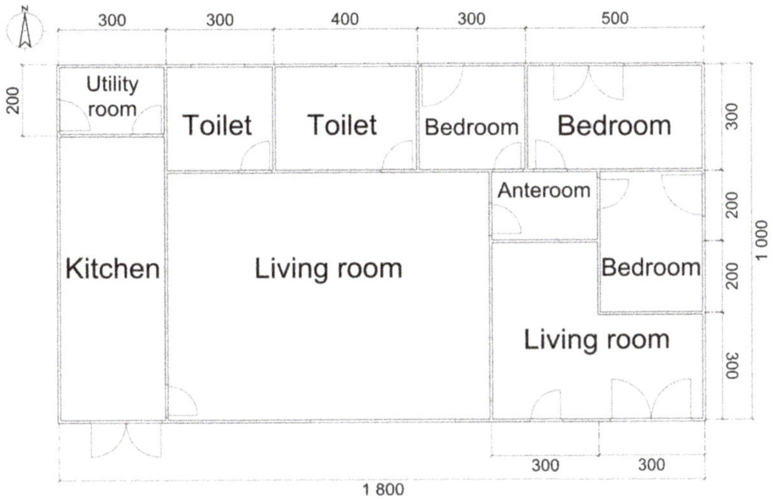

Figure 1. Plan views of the benchmark single detached house (dimensions in cm).

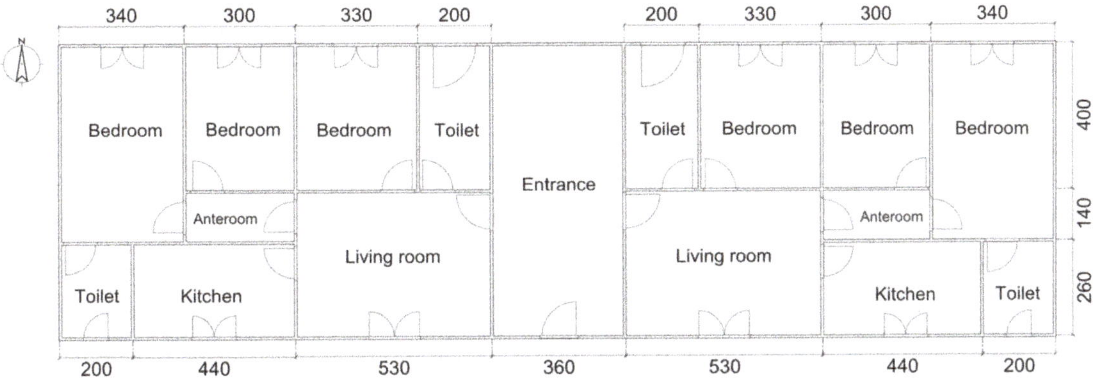

Figure 2. Plan views of one storey of the benchmark apartment block (dimensions in cm).

2.1.2. Assessment of Thermal Needs

The thermal needs (heating, cooling, DHW) of the two benchmark buildings were assessed considering three different levels of insulation (original building, partial renovation, total renovation of the building envelope) and two climatic conditions (Torino and Oulx). Therefore, 12 energy simulations were performed using the computer suite developed by ANIT [35], and they are hereby indicated as scenarios S01–S12. The ANIT package consists of several programs, among which the main one is called LETO. LETO is based on the Italian standard series UNI/TS 11300 [36], i.e., the national implementation of the international standard ISO 13790:2008 [37] and successive/related norms. The UNI/TS 11300 is prescribed by Italian law for the energy certification of every new built, refurbished, sold, or rented housing unit or building. The calculation of thermal losses from the building envelope uses monthly mean outdoor temperatures, whereas the peak thermal loads are calculated steady state using the design temperatures (minimum for heating, maximum for cooling) prescribed by the related norm UNI 10349:2016 [38]. Part 4 of the UNI/TS 11300 [36] prescribes the use of the temperature bin method to calculate the COP and EER of heat pumps and air conditioners.

Table 2 reports the yearly heating and cooling needs calculated with the software LETO for the 12 scenarios. The results of the energy simulations show that a reduction of 9–12% in heating loads is achieved with a partial renovation; instead, a more substantial reduction (73–85%) is obtained with a refurbishment of the whole envelope. Climate conditions also have an influential role, as the heating needs in Oulx are up to 58% higher than those in Turin. As expected, this percentage is in line with the difference in heating degree-days according to DPR 412/93 [34] (+56.7% in Oulx, i.e., 4100 vs. 2617), with direct consequences in the monthly energy demand (see Section 2.1 of the Supplementary Materials). On the other hand, the improvement in the building thermal insulation results in an increase in cooling needs up to +146% in the case of a total renovation of the envelope. Cooling needs are typically low in Piedmont and, not surprisingly, only 13.3% of residential buildings in this region currently have a cooling system installed [39]. Hence, our analysis first focused only on heating and DHW. Cooling loads were subsequently included, and the analysis was limited to the completely renovated buildings (S09–12) since, as shown in Table 2, they are the only ones where cooling demand is relevant.

Table 2. Heating and cooling demand (MWh/year) and, between brackets and in italics, the demand per unit heated/cooled area (kWh/m^2/year) of the modelled buildings.

Model Building	Location	Original Building			Partial Renovation			Complete Renovation		
		#	Heating	Cooling	#	Heating	Cooling	#	Heating	Cooling
Single detached house	Turin	S01	29.05 *(186.22)*	0.87 *(5.58)*	S05	26.42 *(169.36)*	0.83 *(5.32)*	S09	7.85 *(50.32)*	2.34 *(15.00)*
	Oulx	S02	42.61 *(273.14)*	0.00 *(0.00)*	S06	38.62 *(247.56)*	0.00 *(0.00)*	S10	10.96 *(70.26)*	0.51 *(3.27)*
Apartment block	Turin	S03	115.37 *(123.52)*	12.32 *(13.19)*	S07	101.25 *(108.40)*	12.53 *(13.42)*	S11	17.03 *(18.23)*	24.81 *(26.56)*
	Oulx	S04	172.89 *(185.11)*	0.68 *(0.73)*	S08	152.35 *(163.12)*	0.89 *(0.95)*	S12	26.91 *(28.81)*	13.14 *(14.07)*

2.2. Heating and Cooling Plant Sizing

Thermal needs calculated in the different buildings and climate conditions are the input for the sizing of heating, cooling and DHW production plants. Seven technologies were considered for the heating and DHW production system: a condensing boiler powered by fossil fuels (natural gas, LPG, or diesel oil) or wood biomass (wood log or pellet), and two types of heat pumps (air source and ground source). Heat pumps can cover every thermal need with a unique system; on the other hand, split air-source chillers were assumed to provide cooling in cases where fossil fuel and wood biomass boilers were installed for heating and DHW production.

The size of every heating and cooling generator was determined based on the maximum thermal loads resulting from simulations. In particular, the peak load is calculated in steady state, imposing a design temperature calculated by the software LETO using the standard UNI 10,349:2016 [38]. The resulting design temperatures for heating are −8 °C for Turin (scenarios S01, S03, S05, S07, S09, S11) and −12.84 °C for Oulx (scenarios S02, S04, S06, S08, S10, S12).

Although the peak heating power requested was lower, such as in the well-insulated case studies, the minimum capacity considered for fossil fuel and biomass boilers was about 30 kW (with slight differences depending on the brand and model) as it is the entry-level power for most manufacturers and it permits production of DHW on demand (i.e., without storage tank). On the other hand, the heat pump units were sized according to the effective values of peak heating and cooling demand.

For ground-source heat pumps, it is necessary to size the Borehole Heat Exchangers (BHEs) as well. For this purpose, the software Earth Energy Designer (EED) [40] was used, which is based on the Eskilson subsurface heat transport model [41]. Three different values of ground thermal conductivity (low: 1.6 $Wm^{-1}K^{-1}$, medium: 2.4 $Wm^{-1}K^{-1}$, high: 3.2 $Wm^{-1}K^{-1}$) were hypothesized to cover the range of most likely values [42]. A minimum operating temperature was set to −3 °C, assuming that the BHE runs with propylene glycol 25% vol. (freezing point: −10 °C) as a heat carrier fluid [43]. Supplementary Materials (Section 2.3) report further details on the sizing procedure and results. In Supplementary Materials Section S.2.3, the area occupied by the BHE field is reported as well, and this is a key figure to assess the feasibility of the shallow geothermal solution, especially in urban areas.

The benchmark buildings considered in our analysis use radiators as heating terminals. When a heat pump was hypothesized, we assessed whether existing radiators could cover the peak heating demand with an operating temperature reduced from the typical value used for boilers (average inlet–outlet 70 °C) to a lower value (inlet at 45 °C) suitable for single-stage heat pumps. The thermal power provided by a radiator is described by the relation

$$P(\Delta T) = a \cdot \Delta T^{1.3} \tag{1}$$

where a ($W{\cdot}K^{-1.3}$) is a constant that depends on the radiator size and ΔT (K) is the temperature difference between the average water temperature in the radiator and the ambient temperature (set to 20 °C). The nominal power of radiators P_{nom} is provided for $\Delta T = 50$ °C, i.e., for an average radiator temperature of 70 °C. The maximum delivery temperature for radiators working with heat pumps was set in this study to 45 °C (with the return at 40 °C) to guarantee a sufficient coefficient of performance (COP). At this operating temperature, the thermal power delivered by the radiator is

$$P(\Delta T) = P_{nom} \cdot \left(\frac{\Delta T}{50}\right)^{1.3} \tag{2}$$

Therefore, with the maximum delivery and return temperatures considered (45 °C and 40°, respectively, i.e., $\Delta T = 22.5$ °C), the nominal thermal power of radiators that is needed to cover the peak demand P_{max} is

$$P_{nom} = \left(\frac{50}{22.5}\right)^{1.3} \cdot P_{max} = 2.824 \cdot P_{max} \tag{3}$$

For each scenario, we checked if the nominal power installed was sufficient to cover the peak thermal demand and, if not, the radiators were replaced with larger ones, and their cost was considered in the economic analysis (see Section 2.3). More details on the sizing of radiators are reported in the Supplementary Materials (Section S2.2).

Of course, radiators cannot provide cooling. Therefore, when the cooling demand was included in the analysis, the existing radiators were replaced with fan coils (and their cost accounted in the LCOH calculations) in the cases of air-source and ground-source heat pumps. In the case of fossil fuel and biomass boilers, additional mono-split air conditioners were installed in each room to provide this service. The size of air conditioners is based on the peak cooling needs derived from the energy simulations.

Coupling a heat pump with radiators needs thermal storage since the working temperature difference of a heat pump is usually equal to 5 °C, whereas radiators operate with larger temperature differences (10–20 °C). The thermal storage tank was sized equal to 20 L per kW of heat pump nominal power (see Supplementary Materials, Section S2.2). In addition, hot water storage (200 L/apartment, considering four residents) is required because DHW heat pumps are not sized to produce hot water instantaneously in the same way as a boiler.

Finally, photovoltaic (PV) panels were sized considering only the actual electricity yearly demand of the thermal plant and not the consumption related to other components, such as electrical appliances. Therefore, PV panels were not simulated in the case of boilers because they require a small amount of electricity just for the ignition and the control systems. Concerning heat pumps and additional split systems, PV panels were instead sized according to both the energy demand and the legislative requirements. In particular, the Legislative Decree 28/2011 imposes a minimum PV power of 1 kW installed per each 50 m^2 of plan view footprint of the building [44]. Therefore, the installed capacity to cope with this requirement is 3.6 kW for the detached house and 4.3 kW for the block of flats. The PV plants were sized according to these thresholds, i.e., to provide at 3.6 kW or 4.3 kW, respectively, for the detached house and the block of apartments. Further details on PV system sizing are reported in Section S2.4 of the Supplementary Materials.

2.3. Estimation of the Initial Investment

The initial investment for the energy refurbishment of buildings was estimated considering only the equipment for heating, cooling and DHW production, without accounting for the other cost items that are equal for all technological configurations assumed. For this reason, the study did not include the expenditure for the (partial or complete) refurbishment of the building envelope. Instead, the cost items considered are the procurement and the installation of the heating, cooling and DHW supply systems, PV panels and, when required, the replacement of radiators.

Each thermal plant required a different approach in determining its price despite using similar references, such as market surveys and catalogues from manufacturers. Only two models of boilers were chosen: a small-capacity one (below 35 kW) for the single detached house (scenarios S05, S06, S09, S10) and all the deeply refurbished block of apartments (scenarios S11, S12), and a more powerful boiler (65–70 kW) for the block of apartments that underwent a partial refurbishment (scenarios S07, S08). The reason lies in the small dependence of boiler price on thermal power, as the data in Table 3 show.

Table 3. Boilers unitary prices considering heater, accessories, and installation cost (excluding VAT) [45–50].

Typology	Cost (<35 kW)	Cost (65–70 kW)
Gas boiler	EUR 6860	EUR 9255
LPG boiler	EUR 6860	EUR 9255
Oil boiler	EUR 7175	EUR 9255
Wood logs boiler	EUR 7015	EUR 9270
Pellet boiler	EUR 7015	EUR 9270

On the other hand, the price of heat pumps strongly depends on their capacity, and a typical approach adopted in the literature is therefore to calibrate a power–cost correlation based on known values [51,52]. As shown in Figure 3, two correlations can be found between the thermal power P (kW) and the cost C (EUR) for air-source and ground-source heat pumps, respectively

$$C_{ASHP} = 1168 \cdot P^{0.8364} \text{ with } P < 1 \text{ MW} \tag{4}$$

$$C_{GSHP} = 2485 \cdot P^{0.6094} \text{ with } P < 1.7 \text{ MW} \tag{5}$$

Finally, a linear capacity–cost correlation was found for mono-split air conditioners, based on the regional price list [53]

$$C_{Mono\ split} = 82.70 + 361.49 \cdot P \text{ with } P < 7 \text{ kW} \tag{6}$$

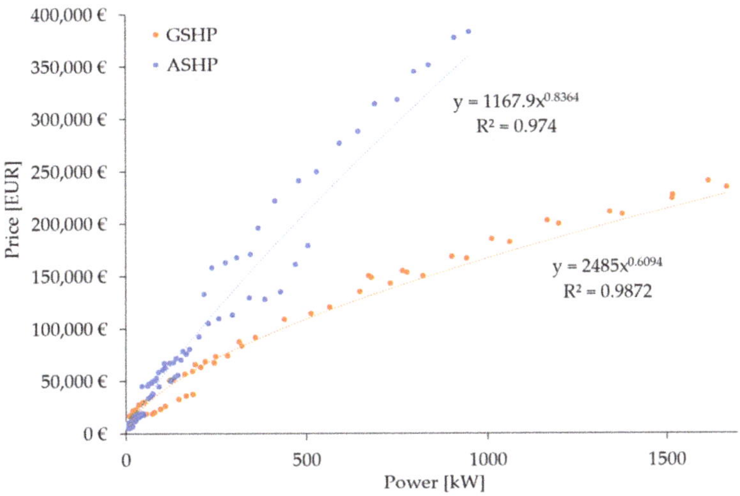

Figure 3. Correlations between capacity and cost for air- and ground-source heat pumps (elaboration from Refs. [54,55]).

The drilling and installation cost of BHEs was set to 50 EUR/m, VAT excluded, based on personal communications from Italian practitioners. This value lies in the range of other literature sources, such as references [56,57].

As for the purchase and installation costs of the PV plant, the price of PV panels is 750 EUR/kW, and the price of a single-phase inverter (5 kW) is EUR 980. The installation cost was estimated at 148.58 EUR/m^2, based on regional price lists. Considering that PV panels occupy 5.45 m^2/kW, the purchase and installation of PV panels imply a total cost of 1755 EUR/kW (excluding value-added tax—VAT).

In the cases where radiators needed to be replaced, steel radiators were considered, with a cost of 130 EUR/kW of nominal power (P_{nom}) calculated with Equation (3).

As mentioned above, the replacement of radiators with fan coils was hypothesized only if cooling was considered. A linear correlation of their cost with the heating capacity (valid from 0 kW to 17 kW) was found

$$C_{\text{Fan coil}} = 132.19 + 32.73 \cdot P \text{ for } P < 17 \text{ kW} \quad (7)$$

In this case, the reference temperature for thermal power is 45 °C in heating mode and 7 °C in cooling mode. These temperatures were assumed as the working points in the energy simulations.

2.4. Estimation of Operation and Maintenance Costs

The operational costs were assessed by estimating the quantity of fuel (for boilers) and electricity from the grid (for heat pumps) needed to cover the heating and cooling demand (if any), adopting the unit costs available in the literature.

The maintenance costs were set considering the different actions that needed to be periodically taken, depending on the technology, and assigning the costs of such actions based on a market inquiry.

2.4.1. Purchase of Fuel and Electricity

The fuel costs for the different boilers considered in this study were calculated considering the unit costs reported in Table 4, which were derived from European and local surveys [21,58,59].

Table 4. Unit prices of energy sources.

Energy Source	Value	Unit	Validity Range Min	Max	Unit	Reference
Natural gas for household consumers	148	EUR/MWh	-	6	MWh	[21]
	93	EUR/MWh	6	56	MWh	
	74	EUR/MWh	56	-	MWh	
Diesel oil	1.34	EUR/l	-	-	-	[58]
	135	EUR/MWh	-	-	-	
LPG (tank on loan)	1.53	EUR/l	-	-	-	[58]
	230	EUR/MWh	-	-	-	
Wood logs	150	EUR/ton	-	-	-	[58]
	41	EUR/MWh	-	-	-	
Pellet (15 kg bags)	290	EUR/ton	-	-	-	[58]
	58	EUR/MWh	-	-	-	
Electricity for household consumers	252	EUR/MWh	1	2.5	MWh	[21]
	234	EUR/MWh	2.5	5	MWh	
	232	EUR/MWh	5	15	MWh	
	225	EUR/MWh	15	-	MWh	
National price of electricity with on-site power exchange	67	EUR/MWh	-	-	-	[59]

The fuel demand (FD_h) for boilers was calculated using the heating demand (Q_h) resulting from the energy simulations (Sections 2.1 and 2.2) and hypothesizing fixed efficiency values for heat generation, distribution, regulation, and emission.

$$FD_h = \frac{Q_h}{\eta \cdot LHV} \quad (8)$$

where Q_h (kWh/year) is the heating demand, η (dimensionless) is the efficiency of the boiler and LHV is the lower heating value of the fuel (kWh/m^3 for gas, kWh/L for LPG and heating oil, and kWh/kg for wood logs and pellet).

The electricity demand (ED_h) for the heat pump in heating mode was derived with the following formula

$$ED_h = \frac{Q_h}{COP} \quad (9)$$

where the coefficient of performance (COP) was calculated referring to catalogues including several combinations of source and supply temperatures [60]. These input data allowed developing COP maps such as those shown for ground-source (Figure 4A) and air-source (Figure 4B) heat pumps. As for the air-source type, Figure 4B shows COP values at full and 50% of nominal power. This representation highlights the positive effect of the inverter on the heat pump performance. In the coldest climate zone (Oulx), only the use of an inverter driven ASHP allows the minimum Seasonal Performance Factor value (SPF = 2.243) set by the Italian legislation to be coped with in order to consider the heat pump a renewable heat source. The nominal power of the ASHP is always greater than for the GSHP due to the need to ensure a sufficient thermal power and COP, both of them diminishing as the outdoor temperature diminishes. The difference, for each scenario, between the nominal power of the ASHP and GSHP is moderate in the detached house cases, whereas this relative difference can be as high as 45% for the larger capacities needed by the block of flats (i.e., above 40 kW). The reason is that the minimum size of ASHPs available on the market often exceeds the maximum thermal power needed by the detached house configurations.

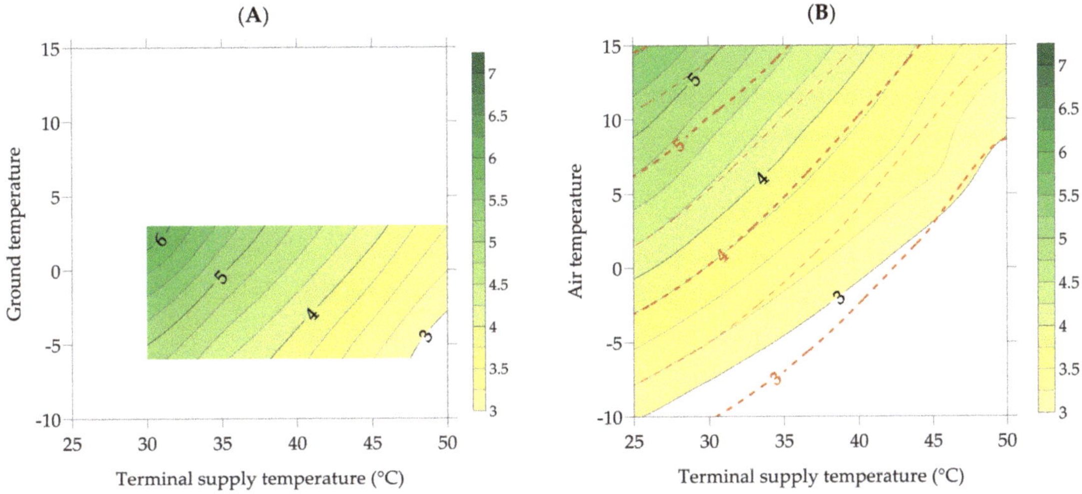

Figure 4. Examples of COP maps of heat pumps (own elaboration with data from [60]): (**A**) GSHP, nominal power 40.3 kW; (**B**) ASHP, nominal power 78.6 kW at full power (black full lines + gradient map) and at 50% power (orange dot lines).

Maps like those reported in Figure 4 were developed for the energy efficiency ratio (EER) and the electrical demand (ED_c) to cover the cooling demand Q_c was calculated with the following formula

$$ED_c = \frac{Q_c}{EER} \tag{10}$$

Both the values of COP in Equation (9) and of EER in Equation (10) are calculated using the air temperature bins in the case of the use of an ASHP and split air conditioners, or the monthly average fluid temperatures calculated with the software EED (see Section 2.3 and, in the Supplementary Materials, Section S2.3), in the case where a GSHP is adopted.

2.4.2. PV Systems and On-Site Exchange

In the case of heat pump and when installing additional splits for cooling, a PV system covers the electrical demand to reduce operational costs as much as possible. Each nominal kW installed was assumed to produce 1.1 MWh/year, i.e., a cautious hypothesis for a PV system installed in Piedmont.

The annual electricity expenditure was estimated considering the adoption of the "on-site exchange" mechanism managed by the national energy services authority [61]. With this mechanism, the self-consumed energy is free-of-charge, whereas that drawn from the grid is paid at the conventional fare (see Table 4). Finally, the amount delivered to the grid is paid by GSE at lower rates (from 90 EUR/MWh to 150 EUR/MWh) than those reported in Table 4 since taxes and invoices are not paid back. The rate for each case study was calculated according to the national guidelines [61].

The on-site exchange mechanism rewards self-consumption instead of delivering to the grid and, hence, thermal and electricity storage was assumed, aiming to reduce the electricity exchange with the grid as much as possible. Storage batteries were therefore hypothesized for accumulating the electricity produced by the PV panels during the day and not self-consumed instantaneously by the thermal plant. The sizing of the capacity (E_{bat}, kWh) depends on the ratio between the energy that would be delivered to the grid without batteries installed (E_{del}, kWh) and the energy consumed by the thermal plant when the PV panels are off (E_{off}, kWh). The battery capacity was calculated with the following relation

$$E_{bat} = \min\left(\frac{\sum_{i=1}^{365} E_{del,i}}{365}, 0.8 \cdot \frac{\sum_{i=1}^{365} E_{off,i}}{365}\right) \quad (11)$$

2.4.3. Maintenance Costs

Maintenance costs were evaluated considering both the operations that are compulsory by law (flue gas analyses every two years for boilers with solid and liquid fuel, and every four years for gaseous fuels) and those advised by boiler and heat pump manufacturers (e.g., refrigerant leakage control and possible replacement). The yearly maintenance costs (Table 5) were estimated integrating data from reference [62] and a market inquiry, finding a narrow difference between different heating and cooling technologies.

Table 5. Maintenance costs for boilers and heat pumps.

Intervention	Capacity Range (kW)	Cost (EUR, VAT Excl.)
Preventive maintenance for natural gas boilers	<35 kW 35–60 kW 60–100 kW	80 120 150
Preventive maintenance for oil boilers	<35 kW 35–60 kW 60–100 kW	110 130 180
Preventive maintenance for wood boilers	<35 kW 35–100 kW	150 250
Preventive maintenance for pellet boilers	<35 kW 35–100 kW	110 220
Combustion analysis	<35 kW 35–100 kW	40 50
Descaling of exchangers and boilers	<35 kW	50
Heat pumps and chillers	<35 kW 35–100 kW	150 250

As for the replacement of parts of the heating/cooling plant, the lifetime length of boilers, PV panels and heat pumps is supposed to be equal to 20 years at least, i.e., no replacement occurs for 20 years. While this is a straightforward assumption for boilers and PV panels, the operating lifetime of heat pumps is more debated [63–65]. In particular, the compressor of the heat pump is more prone to wearing and, hence, a replacement was considered in the 11th year of operation, i.e., after ten years of operation. The cost of such a replacement was estimated based on a market inquiry, and the following relation was found

$$C_{compressor} = 707.74 + 105.73 \cdot P \quad (12)$$

where P (kW) is the thermal power of the heat pump.

The replacement of the compressor was also considered for the split air conditioners installed to cover the cooling demand in a boiler-based configuration. This assumption is conservative since these plants operate for a limited number of hours per year (i.e., much less than heat pumps do), and hence the compressor is likely to last for the whole 20 year period.

As for the PV system, the panels are deemed to last well beyond 20 years, but the inverter is generally replaced every 10 years [66,67]. An inverter replacement was therefore considered in the 11th year of operation, and its cost was derived from a few data on a regional price list [53] as directly proportional to the plant power P (kW)

$$C_{inverter} = 196.05 \cdot P \quad (13)$$

2.5. Levelized Cost of Heat (LCOH)

The economic viability of different technologies for heating, cooling and DHW production was evaluated using the Levelized Cost of Heat (LCOH) during a lifetime of 20 years [62]. This period is short enough to assume that boilers, heat pumps and PV panels will not need to be entirely replaced [68].

The Levelized Cost of Heat $LCOH$ (EUR/kWh) is defined by the following equation

$$LCOH = \frac{\sum_{t=1}^{n} \frac{I_t - R_t + F_t}{(1+i)^t}}{\sum_{t=1}^{n} \frac{E_t}{(1+i)^t}} \tag{14}$$

where I_t (EUR) is the investment cost incurred in the year t, R_t (EUR) is the amount eventually refunded by incentives in year t (EUR), F_t (EUR) is the annual expenditure for energy sources, E_t (kWh) is the energy delivered by heating terminals directly to users, $n = 20$ is the lifetime of the system and I is the discount rate.

The discount rate I was assumed to be equal to 4%, a value that falls within the range of those adopted by several studies on energy and buildings [10,69–72]. However, the personal choices and needs of the customer and the economic conjuncture strongly affect the discount rate, which was therefore considered one of the critical parameters for the sensitivity analysis.

As for refunds (R_t), the Italian Ecobonus was considered, which consists of the reimbursement of 65% of the capital expense for the energy refurbishment of buildings, delivered in ten yearly equal payments.

3. Results

The presentation and discussion of the results of this study are divided into three sub-sections with three different insights levels. The first analysis, presented in Section 3.1, was performed on buildings without a cooling system (i.e., heating-only systems). Subsequently, cooling loads were included in the calculations, and the results are shown in Section 3.2. Finally, the limitations of our study are addressed in Section 3.3: the variability of some major cost items is evaluated with Monte Carlo simulation, and temporary factors influencing the LCOH values that are not considered by our study are described, drawing some qualitative conclusions on their influence.

3.1. Heating-Only Systems

The values of LCOH are reported in Table 6 for the two buildings (detached house and block of flats), the two levels of insulation (partial and total renovation of the envelope) and the two climate zones (Turin and Oulx). The values are calculated in the presence of incentives whose contribution to the reduction in the LCOH is shown.

The data reported highlight that there is generally a clear gap of LCOH values between renewable and fossil fuel heating technologies when the Ecobonus incentive is applied. The most expensive renewable heating technology (GSHP with the lowly conductive ground in the scenarios S06 and S08; pellet in the scenarios S05, S09, S10, S11, S12; ASHP in the scenario S07) is always cheaper than a natural gas boiler (with a gap up to 33.6%), except for the scenario S08 (partially renovated apartment block in Oulx) where the GSHP with lowly conductive ground is slightly more expensive than the gas boiler due to the very high costs for BHE drilling. The other two fossil fuels (LPG and heating oil) have an even wider gap: that is, renewable heating technologies allow an LCOH reduction of about 60–70% compared to LPG and 40–50% compared to heating oil.

Table 6. LCOH values (EUR/MWh) for the different heating-only systems. The LCOH with the application of incentives is reported in the first row of each case, whereas the relative reduction with respect to the scenario without incentives is in the second row. Legend: NG = natural gas condensing boiler, LPG = LPG condensing boiler, OIL = diesel oil condensing boiler, WL = wood logs boiler, PEL = pellet boiler, AS = air-source heat pump, GS = geothermal heat pump (subscripts stand for low, medium and high conductivity, respectively).

Building	Renovation	Location (Scenario)	NG	LPG	OIL	WL	PEL	AS	GS_{lc}	GS_{mc}	GS_{hc}
Single detached house	Partial	Turin (S05)	128.5 −8%	290.7 −4%	179.6 −6%	70.3 −14%	89.6 −14%	78.8 −47%	83.8 −48%	80.7 −48%	78.7 −48%
		Oulx (S06)	127.2 −6%	296.6 −2%	180.2 −4%	62.1 −11%	84.1 −11%	72.6 −46%	84.5 −47%	75.4 −50%	73.2 −49%
	Complete	Turin (S09)	142.2 −18%	292.6 −9%	198.6 −16%	97.7 −24%	118.7 −26%	96.2 −48%	114.6 −50%	111.3 −50%	109.4 −50%
		Oulx (S10)	130.7 −15%	278.3 −8%	183.5 −13%	83.8 −22%	103.3 −23%	58.5 −44%	82.2 −49%	76.2 −49%	74.0 −48%
Apartment block	Partial	Turin (S07)	92.5 −4%	273.9 −1%	163.9 −2%	55.7 −5%	72.1 −4%	81.9 −38%	71.7 −49%	64.4 −49%	61.3 −49%
		Oulx (S08)	91.2 −3%	273.3 −1%	162.7 −1%	53.6 −3%	70.3 −3%	78.7 −31%	101.5 −45%	80.6 −43%	73.2 −41%
	Complete	Turin (S11)	110.9 −8%	252.5 −4%	157.8 −7%	60.3 −14%	77.2 −15%	65.4 −48%	66.6 −49%	63.5 −49%	61.6 −49%
		Oulx (S12)	109.8 −6%	252.9 −3%	155.9 −5%	57.1 −11%	73.7 −12%	62.0 −47%	71.1 −49%	66.6 −48%	64.4 −48%

A conductive ground requires fewer BHEs to be installed, thus reducing the initial investment. Hence, the LCOH of GSHPs decreases as the thermal conductivity of the subsurface increases. Different impacts, however, are observed: while a slight LCOH increase occurs for detached houses, from 5.3% to 19.4%, the increment is substantial (up to 39.9%) for an apartment block. This fact can be explained considering two factors: (i) the mutual interference between probes in large BHE fields makes the borehole length increase hyper-linearly with the thermal power, and (ii) the drilling and installation of BHEs have practically no economy of scale, contrary to the other parts of a GSHP system; therefore, the cost of BHEs accounts for an increasing share of the overall installation costs (up to 60%) as the plant size increases. This result agrees with the literature results, such as reference [43].

The share of LCOH reduction reported in Table 6 highlights the fact that incentives have a strong impact on the economic viability of heat pumps, reducing their LCOH values between 31% and 49%. This result is explained by the fact that, contrary to feed-in tariffs or fixed incentivization, the Ecobonus refunds 65% of the initial investment, thus improving the economic viability of technologies with high installation costs but low operational costs. This effect is expected to be even stronger with the recently introduced Superbonus (110% tax refund of installation costs, see reference [33]), which was not considered in this analysis due to its temporary nature.

3.2. Heating and Cooling Systems

The analysis reported in the previous paragraph deals with heating-only configurations that, as already stated, are quite common in Northern Italy due to its climate [39]; however, the request for cooling systems is increasing due to the higher cooling loads of new, better-insulated buildings (as also shown in Table 2). For this reason, the analysis was extended to cover the cooling demand, and the LCOH was re-calculated combining heating, DHW and cooling. Only the case studies with complete renovation were considered since they involve the most relevant space cooling demand.

The inclusion of a cooling service depends on the different heating technologies adopted and the resulting supplying mode:

- The reversible heat pump models chosen in the heating-only analysis have a sufficient size to cover the cooling demand (except for scenario S11—the completely renovated apartment block in Turin—which requires an additional power of just 3 kW). However, including the cooling service makes it necessary to replace all radiators with fan coils (see Section 2.3).

- For biomass and fossil fuel boilers, the cooling needs were deemed to be covered by a few mono-split air conditioners and, hence, a modest additional investment is needed.

When space cooling is included in the economic analysis, this results in an increase or a decrease in LCOH values, as shown in Table 7. Indeed, the initial investment is shared on a larger quantity of heat (delivered to or removed from the building), and, in some cases, cooling has a lower unit cost compared to heating. In these cases, including the cooling service leads to a reduction in LCOH values. GSHPs noticeably benefit from the inclusion of the cooling demand because this reduces the heat imbalance underground, thus leading to a lower overall BHE depth to drill. This effect is particularly evident for the block of flats due to the above-mentioned BHE mutual interference. There are, however, several cases where the inclusion of cooling increases the overall LCOH. The reason lies in a combination of three possible factors, namely (i) low demand for cooling, as in the cold climate zone (Oulx), (ii) a lower LCOH for heating than for cooling, as is the case of biomass boilers and (iii) the need to replace radiators with fan coils to cover a modest cooling demand (e.g., for scenario S10).

Table 7. LCOH values (EUR/MWh) with incentives referred to heating and cooling solutions and the relative variation compared to the heating-only case ($\Delta_{h\&c}$). Legend: NG = natural gas condensing boiler, LPG = LPG condensing boiler, OIL = diesel oil condensing boiler, WL = wood logs boiler, PEL = pellet boiler, AS = air-source heat pump, GS = geothermal heat pump (subscripts stand for low, medium, and high conductivity, respectively).

Building	Location	Quantity	NG	LPG	OIL	WL	PEL	AS	GS_{lc}	GS_{mc}	GS_{hc}
Single detached house	Turin (S09)	LCOH	171.3	293.4	217	134.4	155	93.1	110.8	108.4	107.1
		$\Delta_{h\&c}$	20%	0%	9%	38%	31%	−3%	−3%	−3%	−2%
	Oulx (S10)	LCOH	176.2	318.4	227.1	130.7	152.5	88.5	108.6	102.9	100.8
		$\Delta_{h\&c}$	35%	14%	24%	56%	48%	51%	32%	35%	36%
Apartment block	Turin (S11)	LCOH	106.8	184.8	132	78.1	89.8	56	52.6	48.7	47.1
		$\Delta_{h\&c}$	−4%	−27%	−16%	30%	16%	−14%	−21%	−23%	−24%
	Oulx (S12)	LCOH	114.8	223.8	149.4	74.2	88.7	60.6	54.5	52.4	50.7
		$\Delta_{h\&c}$	5%	−12%	−4%	30%	20%	−2%	−23%	−21%	−21%

3.3. *Study Limitations*

3.3.1. Uncertainty on Cost Items: Sensitivity Analysis

The LCOH is an easily understandable indicator and makes the comparison among different technologies very immediate despite some inevitable simplifications [10]. A major limitation is represented by the variability in some installation cost components, such as the unit cost of BHE drilling, which depends on the geological characteristics of the ground and the local, nearly monopolistic market [73,74]. For this reason, the estimation of LCOH requires the sensitivity analysis to be more rigorous [5,62].

We identified the prices of energy sources, heat pumps, BHEs and batteries, and the discount rate as the most sensitive parameters and, therefore, as sources of uncertainty. The sensitivity analysis focused on verifying how LCOH varies according to those parameters. The study was performed through a Monte Carlo simulation hypothesizing that variables randomly range between a minimum and a maximum value (Table 8).

Table 8. Minimum and maximum values assumed in the sensitivity analysis.

Variable	Unit	Min	Max	Notes
Electricity price for household consumers	EUR/MWh	49	334	1000–2500 kWh
	EUR/MWh	98	301	2500–5000 kWh
	EUR/MWh	95	280	5000–15,000 kWh
	EUR/MWh	95	313	>15,000 kWh
National price on power exchange	EUR/MWh	0	163	-
Natural gas price for household consumers	EUR/MWh	31	160	<20 GJ
	EUR/MWh	28	107	20–200 GJ
	EUR/MWh	27	104	>200 GJ
Diesel oil	EUR/l	1.06	1.51	-
LPG	EUR/l	1.45	1.62	-
Wood	EUR/kg	0.13	0.18	-
Pellet	EUR/kg	0.24	0.34	-
BHEs drilling and setting cost	EUR/m	40	75	-
ASHP cost	Variation	−9%	+9%	-
GSHP cost	Variation	−9%	+9%	-
Batteries cost	Variation	−12%	+12%	-
Discount rate	%	2%	6%	-

As the Italian prices for the energy sources were assumed in our study, the sensitivity analysis considered a possible variation in their values with different assumptions. As for electricity and natural gas, the adopted range for the sensitivity analysis is between the lowest and the highest unit cost in Europe for every consumption band by Eurostat [21]; the lowest and the highest electricity prices are usually in Bulgaria and Germany, respectively. Instead, the unit prices of natural gas for different demand ranges are quite different depending on the country [21]. The variations in heating oil, LPG and wood biomass prices were hypothesized based on the surveys published by the Chamber of Commerce of Turin in the years 2017–2020 [58]. The reference for the maximum BHE cost is the regional price list [53]; the minimum cost is the result of a survey performed by the authors among workers in the sector. The minimum heat pump cost was computed considering that the price may decrease by 9% until 2030 (a 9% annual drop) [75,76]. Symmetrically, the maximum value accounted for a 9% increase in the capital cost. In the same way, battery prices ranged from −12% to +12% [77]. Heat pumps and batteries were considered sources of uncertainty because they are less consolidated technology compared to boilers and require critical materials, such as rare-earth metals. Finally, the discount rate varied between 2% and 6% based on the work of Cui et al. [10].

Figure 5 reports the sensitivity analysis results for the single detached house in Turin (all the other cases are in Section S3 of the Supplementary Materials). The plots show that diesel oil and LPG boilers are usually the two most expensive solutions under every hypothesis of random price combination. The LCOH values for other energy sources, especially wood log boilers, exhibit a low variability from its median. Only the curve of gas boilers has a steeper slope due to the gas prices' fluctuation. Gas boilers are usually more expensive than ASHP and wood boilers, particularly when incentives apply. GSHPs also turn out to be cheaper than gas boilers only in the presence of incentivization; on the other hand, gas boilers are usually cheaper than GSHPs. Although it considered wide ranges for input parameter values, the sensitivity analysis showed a ranking of LCOH values such as those presented in Sections 3.1 and 3.2 with the default parameter values.

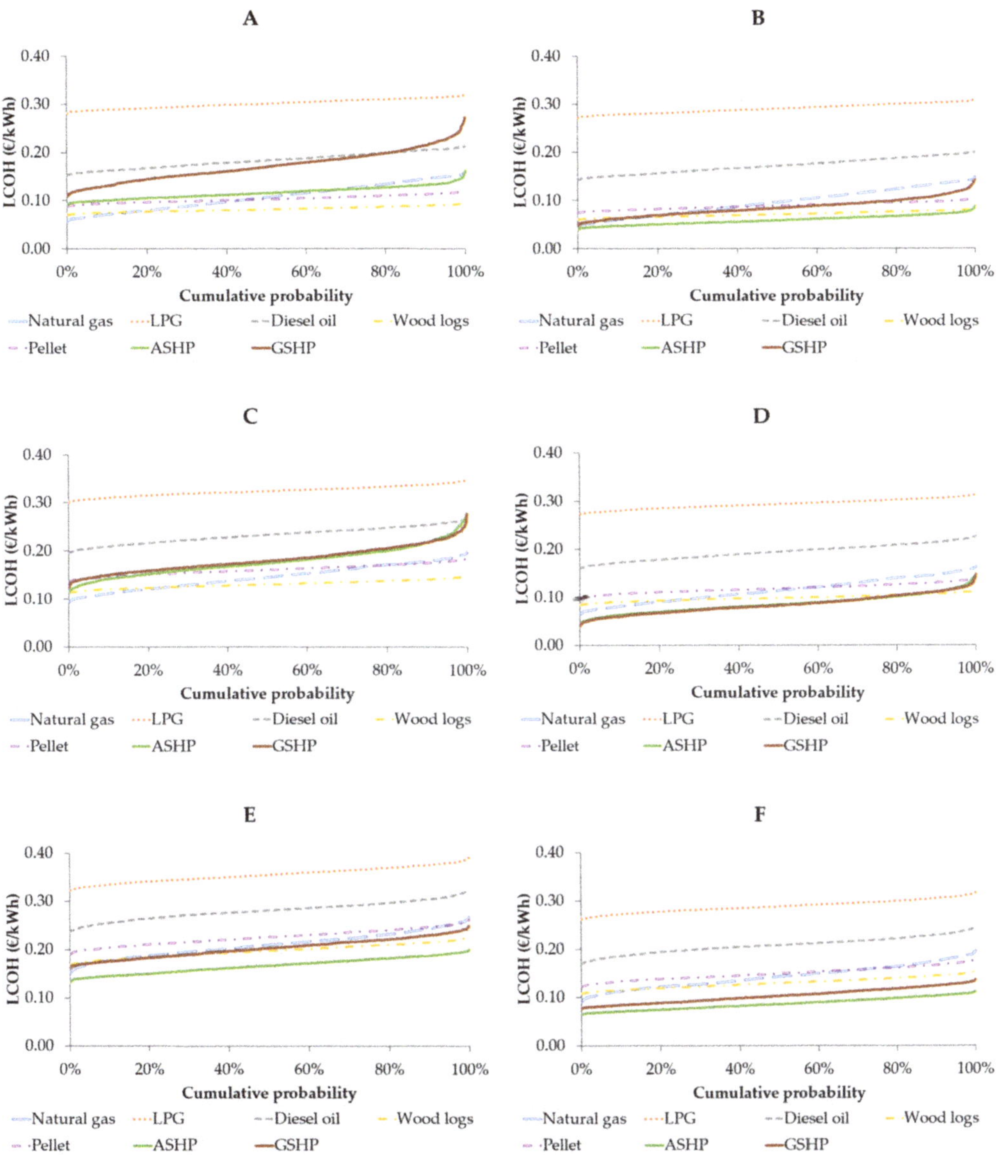

Figure 5. Probabilistic distributions of LCOH values in a detached house in Turin in the following cases: (**A**) heating-only after partial renovation without incentives; (**B**) heating-only after partial renovation with incentives; (**C**) heating-only after complete renovation without incentives; (**D**) heating-only after complete renovation with incentives; (**E**) heating and cooling after complete renovation without incentives; (**F**) heating and cooling after complete renovation with incentives.

3.3.2. Temporary Factors Influencing LCOH

Our study did not consider two temporary factors, namely (i) the introduction of Superbonus in 2020 [33] and (ii) the energy price increase occurring since late 2021 and recently exacerbated by the war in Ukraine.

As stated above, the Superbonus is the refunding of 110% of the initial investment for the energy refurbishment of a building, which is provided in the form of a tax deduction (5 yearly payments after the conclusion of the building restoration). This tax deduction can be transferred to a bank, thus receiving an immediate refund of about 95% of the installation costs incurred. As this incentive virtually nullifies the upfront costs of a building refurbishment intervention, the LCOH is significantly reduced for heat pumps, which require the greatest initial investment. The Superbonus, therefore, represents a strong incentive to adopt heat pump technologies and, to a lesser extent, biomass boilers. However, this is expected to last for a short time: at present, its phase-out is foreseen for December 2022 for detached houses and for December 2023 for other buildings.

Since the second half of 2021, the price of fossil fuels has been increasing at a fast rate due to the post-COVID pandemic economic rebound, with a consequent rapid increase in the demand, and to several supply reduction issues [78]. At the beginning of 2022, with the increasing tensions culminating with the Russian invasion of Ukraine, gas prices increased to unprecedented price levels: for example, the natural gas EU Dutch TTF price—which never exceeded 40 EUR/MWh between 2010 and 2020—exceeded 200 EUR/MWh in March 2022 [79]. The impact of this energy price shock is currently unpredictable, especially if put into a 20-year perspective. The only partial conclusion that can be drawn is that this price shock moves the focus of an economic viability analysis from the upfront costs to the operation and maintenance costs.

4. Discussion

The results reported above highlight that the Ecobonus makes fossil fuel boilers no longer an economically convenient option for the renovation of buildings. While the margin is still narrow compared to gas boilers, in the case of LPG/heating oil boilers it is so wide that it should stimulate the replacement of all these heating systems with renewable energy technologies in the areas not reached by gas pipelines. This result is consistent with a recent study by Casasso et al. (2019, [23]) in the region of Aosta Valley (NW Italy, bounding with Piedmont). A good agreement is also observed with a previous study of Martinopoulos et al. (2018, [80]) based on slightly older figures (the year 2014) in 16 countries, including Italy. The LCOH values of gas boilers in a detached house were higher compared to the analysis of Wang et al. (2018, [15]) in the UK (i.e., 127.2–142.2 EUR/MWh compared to about 95 EUR/MWh, respectively), whereas the LCOH values of air-source and ground-source heat pumps were lower (respectively: 58.5–96.2 EUR/MWh vs. about 120 EUR/MWh; 75.4–111.3 EUR/MWh vs. about 135 EUR/MWh). As shown later, this discrepancy is explained by the differences in electricity and gas costs in Italy and the UK. Compared to the recent study of Novelli et al. (2021, [81]), the values of LCOH of our study are higher. Nevertheless, the ranking of the economic viability of technologies is the same (GSHPs are slightly more convenient than gas boilers and much better performing than oil boilers), and the effect of the Ecobonus incentive on the LCOH of GSHPs is similar. In addition, the impact of a rooftop PV system on the LCOH of a heat pump appears to be limited in both studies.

The figures on the economic viability of heating technologies can be analyzed considering the diffusion of renewable heating technologies. The coverage of residential heating demand in Italy from 1990 to 2015 has seen a gradual phase-out of oil heating, a noticeable increase in biomass heating in the mid-2000s and the onset of district heating in the 2010–2015 period [82]. More recently, heat pumps have been significantly increasing in Italy. A recent report by the Polytechnic University of Milan estimated an overall investment in 2019 of EUR 1514 M on heat pumps and of only EUR 381 M for condensing boilers [83]. Although heat pumps have overtaken boilers in terms of investments, the absolute numbers of installations are still lower due to the lower costs of boilers compared to heat pumps. However, this result highlights the effectiveness of the Ecobonus in reducing the barrier of the higher upfront cost of heat pumps, which, for example, is still very considerable in the UK [84]. In Italy, the issue of the investment cost has been further addressed by introducing credit transfer of the incentives: the credit owner can transfer it to other legal entities and retrieve almost all the money spent immediately, rather than in 10 yearly quotas [85]. These institutions, such as banks and firms, use the entire credit as their own and generate a margin equal to the percentage not reimbursed to the applicant.

As stated in the introduction, the electricity-to-gas unit cost ratio is a key parameter to drive the electrification of heating through the introduction of heat pumps [20]. Figure 6A shows the trend of this quantity, since 2007, for the European Union (average of 27 EU countries), Italy, the two largest EU economies (Germany and France), the best-performing country from this point of view (Sweden) and the worst-performing country (Belgium). The dashed lines report the cost ratio excluding all taxes and levies from electricity and gas costs. The values of the electricity-to-gas cost ratio of Italy is in line with the EU-27 average and is almost not influenced by taxation. The taxation on electricity is generally higher than for gas, increasing the cost ratio and hindering the diffusion of heat pumps. As shown in Figure 6A, an exception was represented by Sweden up to 2019, where natural gas was more taxed than electricity but, currently, it is slightly lower. Conversely, the German taxation on electricity has soared in the last decade, thus slowing down the increase in the heat pump market. This evidence is confirmed by Figure 6B, which shows a clear inverse relationship between the electricity-to-cost ratio and the turnover of the national heat pump markets (EUR/year per inhabitant, 2019 figure). Both values are, for Italy, very close to the European average.

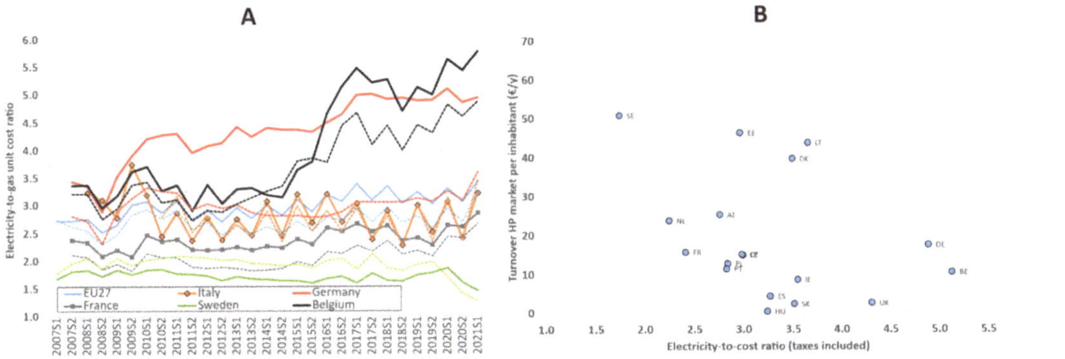

Figure 6. (**A**) The trend of the electricity-to-gas cost ratio in the European Union (EU27), Italy, Germany, France, Sweden and Belgium between 2007 and 2021. The full-line curves refer to the cost ratio, including all taxes and levies, whereas dashed lines refer to the cost ratio without taxes. Data sources: Eurostat database [21], electricity prices for households (500 to 5000 kWh/year), gas prices for households (20 to 200 GJ/year). (**B**) Scatterplot of the electricity-to-gas cost ratio (2019, 2nd semester) vs. the turnover of the heat pump market (the year 2019) normalized on the population (source: EHPA [86]) for 18 European countries.

Statistics on energy consumption in households (Eurostat 2020, [87]) show that biomass heating covers 22.4% of the heating demand in the European Union (EU-27). As shown in Figure 7, the diffusion of biomass heating is correlated with the gas cost expressed as the purchasing power standard (EUR PPP/MWh, source: Eurostat, 2020 [21]). Again, Italy (23.9%) is quite aligned with the EU-27 average figure. Available data do not distinguish between wood logs and pellets; however, specific data on pellet sales are provided by Bioenergy Europe for a few countries [88]. The largest markets for pellet in 2019 were the UK (9 Mton/year), Italy (3.3 Mton/year), Denmark (2.5 Mton/year), Germany (2.3 Mton/year), Sweden (1.8 Mton/year) and France (1.7 Mton/(year). Based on such data, the highest share of heating demand coverage with pellets are observed for Denmark (31.9%) and Sweden (22.8%).

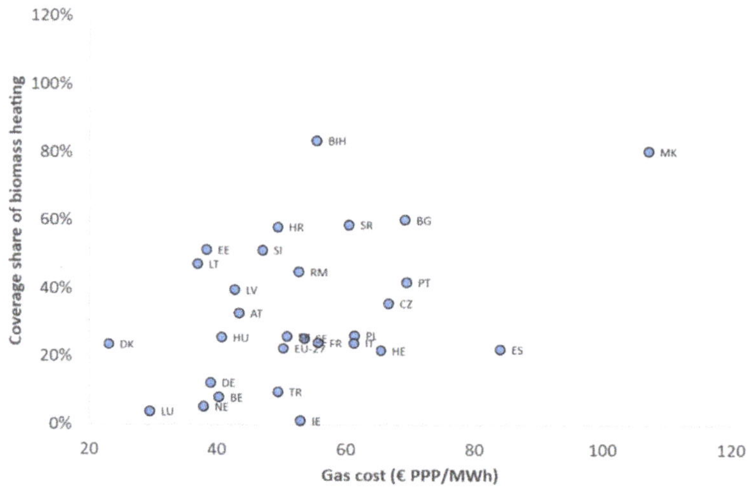

Figure 7. Scatterplot of the coverage share of biomass heating vs. the gas cost (purchasing power parity). Source: Eurostat [21,87].

Based on the diffusion of space heating technologies in the EU, two main conclusions can be drawn, regardless of the incentives in force in each country:

- A low electricity-to-gas cost ratio has a positive effect on the diffusion of heat pumps.
- A high gas cost (compared to purchasing power) has a positive effect on the diffusion of biomass heating.

These facts can be used to shape energy policies to promote the use of heat pumps and, to a lesser extent, biomass heating. As for biomass heating, however, the well-known environmental issues related to wood burning—air pollution, storage space requirements and biomass availability—suggest that the room for further expansion of biomass heating is quite limited, especially in urban areas [89–92].

5. Conclusions and Policy Implications

This study aimed to provide insights into the economic viability of the most common heating and cooling technologies, i.e., fossil fuel boilers (natural gas, diesel oil and LPG), biomass boilers (wood logs and pellet), and heat pumps (both air- and ground-sourced). The analysis was based on an energy simulation of model buildings located in Turin and Oulx, which represent the two main climatic zones of the region, namely the temperate continental and the cold-temperate continental climate.

The installation costs of heating and cooling systems were estimated based on the literature values, regional price lists and catalogues. The LCOH was chosen as the indicator to compare the different technologies. The analysis considered the presence/absence of the Ecobonus incentive and the cooling service. A sensitivity analysis was performed on the most relevant and variable input parameters of the economic analysis.

Overall, the results of this study highlight that, in Italy:

- The Ecobonus incentive regime (and, a fortiori, with the Superbonus) makes renewable energy sources already the most economically viable solutions for space heating and DHW production.
- Wood log boilers are generally the most affordable technology for space heating. Heat pumps (both air-source and ground-source) and pellet boilers follow in this ranking, with variable positions depending on the scenario.
- The impact of the climate zone on LCOH values is generally modest. The only exception is represented by the GSHP systems installed in an apartment block. In this

case, a colder climate (and, consequently, higher heating needs) leads to a relevant increase in the upfront costs of the BHE field.
- Completely refurbished buildings are characterized by a higher LCOH value compared to partially refurbished buildings. This is because life-cycle costs are distributed based on lower heating demand.
- In the climate zones of Piedmont, only completely renovated buildings have a relevant cooling demand.
- The inclusion of space cooling demand makes heat pumps the most economically convenient solution to cover the thermal needs in the residential sector, with a slight advantage of the air-source type for detached houses and of the ground-source type for the blocks of apartments.
- According to the sensitivity analysis carried out, the variability in input parameters of the economic analysis does not substantially alter the ranking of the economic viability of heating technologies.

The results agree with the few studies available in the literature on the LCOH of heating technologies in individual and building centralized plants.

Some policy implications can be drawn from the results of this study and the comparison with energy costs data and heating technologies' breakdown:
- The Ecobonus incentive is a sufficient stimulus to promote renewable heating, at least from the point of view of the LCOH.
- The ratio between electricity and gas unit costs (EUR/MWh) is the critical parameter that drives the diffusion of heat pumps.
- The unit cost of gas (normalized to the purchasing power of the country) is a good predictor of the diffusion of biomass heating.
- Based on the above-reported facts, displacing the taxation from electricity-to-gas and reducing the cost of electricity (e.g., promoting self-production from PV such as in energy communities) can provide a further boost to the growth of heat pump installations.

This study did not deepen the effect of two temporary driving forces that are altering the economic viability of heating technologies: the introduction of the Superbonus and the energy price shock occurring since late 2021. The former virtually nullified the upfront costs of heating and cooling systems, thus benefiting heat pumps; the latter is still difficult to evaluate in the 20-years perspective adopted in the LCOH estimation but seems to be pushing toward a focus on operational costs rather than on the initial investment.

Supplementary Materials: The following supporting information can be downloaded at: https://www.mdpi.com/article/10.3390/en15072375/s1. Section S.1: Introduction; Section S.2: Methods; Section S.2.1: Heating, cooling, and domestic hot water needs; Section S.2.2: Sizing of radiators; Section S.2.3: Sizing of borehole heat exchangers (BHEs); Section S.2.4: Sizing of photovoltaic systems; Section S.2.5: Energy and fuel consumption; Section S.3: Sensitivity analysis.

Author Contributions: Conceptualization, E.R., B.P., A.C. and R.S.; methodology, E.R., B.P., A.C. and R.S.; writing—original draft preparation, E.R. and A.C.; writing—review and editing, E.R., B.P., A.C. and R.S.; supervision, A.C. and R.S. All authors have read and agreed to the published version of the manuscript.

Funding: This research was performed within a contract (number: CIG Z882DAC7B3) of the Polytechnic of Turin with the regional administration of Piedmont—Sustainable energy development sector (Regione Piemonte—Settore sviluppo energetico sostenibile).

Institutional Review Board Statement: Not applicable.

Informed Consent Statement: Not applicable.

Data Availability Statement: Data are contained within the article or supplementary material. Further information is available on request.

Acknowledgments: The authors gratefully acknowledge the regional administration of Piedmont—Regione Piemonte—Sustainable energy development sector (Silvia Riva, responsible for area; Filippo Baretti, Silvia Bonapersona and Giovanni Nuvoli) for their valuable feedback during the research work. The authors also acknowledge the valuable contribution of Sofia Credaro, who assisted in the proofreading and language editing of the manuscript.

Conflicts of Interest: The authors declare no conflict of interest.

References

1. European Commission European Green Deal: Commission Proposes Transformation of EU Economy and Society to Meet Climate Ambitions. Available online: https://bit.ly/36u98ch (accessed on 13 September 2021).
2. European Commission EU Emissions Trading System (EU ETS). Available online: https://ec.europa.eu/clima/policies/ets_en (accessed on 1 October 2021).
3. European Commission Carbon Border Adjustment Mechanism: Questions and Answers. Available online: https://bit.ly/3F7fJpE (accessed on 1 October 2021).
4. European Commission Renewable Energy Directive 2018/2001/EU Known as RED2. Available online: https://bit.ly/RED2directive (accessed on 4 October 2021).
5. Shen, W.; Chen, X.; Qiu, J.; Hayward, J.A.; Sayeef, S.; Osman, P.; Meng, K.; Dong, Z.Y. A Comprehensive Review of Variable Renewable Energy Levelized Cost of Electricity. *Renew. Sustain. Energy Rev.* **2020**, *133*, 110301. [CrossRef]
6. Moseley, P.T.; Garche, J. *Electrochemical Energy Storage for Renewable Sources and Grid Balancing*; Elsevier: Amsterdam, The Netherlands, 2015; ISBN 978-0-444-62616-5.
7. Roser, M. Why Did Renewables Become So Cheap So Fast? Available online: https://bit.ly/3In3Ex0 (accessed on 16 September 2021).
8. Timilsina, G.R. Are Renewable Energy Technologies Cost Competitive for Electricity Generation? *Renew. Energy* **2021**, *180*, 658–672. [CrossRef]
9. Lafond, F.; Bailey, A.G.; Bakker, J.D.; Rebois, D.; Zadourian, R.; McSharry, P.; Farmer, J.D. How Well Do Experience Curves Predict Technological Progress? A Method for Making Distributional Forecasts. *Technol. Forecast. Soc. Change* **2018**, *128*, 104–117. [CrossRef]
10. Cui, Y.; Zhu, J.; Twaha, S.; Chu, J.; Bai, H.; Huang, K.; Chen, X.; Zoras, S.; Soleimani, Z. Techno-Economic Assessment of the Horizontal Geothermal Heat Pump Systems: A Comprehensive Review. *Energy Convers. Manag.* **2019**, *191*, 208–236. [CrossRef]
11. Doračić, B.; Novosel, T.; Pukšec, T.; Duić, N. Evaluation of Excess Heat Utilization in District Heating Systems by Implementing Levelized Cost of Excess Heat. *Energies* **2018**, *11*, 575. [CrossRef]
12. Doračić, B.; Pukšec, T.; Schneider, D.R.; Duić, N. The Effect of Different Parameters of the Excess Heat Source on the Levelized Cost of Excess Heat. *Energy* **2020**, *201*, 117686. [CrossRef]
13. Li, H.; Song, J.; Sun, Q.; Wallin, F.; Zhang, Q. A Dynamic Price Model Based on Levelized Cost for District Heating. *Energy Ecol. Environ.* **2019**, *4*, 15–25. [CrossRef]
14. Vivian, J.; Emmi, G.; Zarrella, A.; Jobard, X.; Pietruschka, D.; De Carli, M. Evaluating the Cost of Heat for End Users in Ultra Low Temperature District Heating Networks with Booster Heat Pumps. *Energy* **2018**, *153*, 788–800. [CrossRef]
15. Wang, Z. Heat Pumps with District Heating for the UK's Domestic Heating: Individual versus District Level. *Energy Procedia* **2018**, *149*, 354–362. [CrossRef]
16. Renaldi, R.; Hall, R.; Jamasb, T.; Roskilly, A.P. Experience Rates of Low-Carbon Domestic Heating Technologies in the United Kingdom. *Energy Policy* **2021**, *156*, 112387. [CrossRef]
17. Connor, P.M.; Xie, L.; Lowes, R.; Britton, J.; Richardson, T. The Development of Renewable Heating Policy in the United Kingdom. *Renew. Energy* **2015**, *75*, 733–744. [CrossRef]
18. Fitó, J.; Dimri, N.; Ramousse, J. Competitiveness of Renewable Energies for Heat Production in Individual Housing: A Multicriteria Assessment in a Low-Carbon Energy Market. *Energy Build.* **2021**, *242*, 110971. [CrossRef]
19. Atalla, T.; Gualdi, S.; Lanza, A. A Global Degree Days Database for Energy-Related Applications. *Energy* **2018**, *143*, 1048–1055. [CrossRef]
20. Bettgenhäuser, K.; Offermann, M.; Boermans, T.; Bosquet, M.; Grözinger, J.; von Manteuffel, B.; Surmeli, N. An Analysis of the Technology's Potential in the Building Sector of Austria, Belgium, Germany, Spain, France, Italy, Sweden and the United Kingdom. Available online: https://bit.ly/3pNv3BL (accessed on 8 March 2022).
21. Eurostat Database—Energy. Available online: https://bit.ly/3klkvr3 (accessed on 17 September 2021).
22. MISE Integrated National Energy and Climate Plan. Available online: https://bit.ly/NECP_IT (accessed on 20 September 2021).
23. Casasso, A.; Capodaglio, P.; Simonetto, F.; Sethi, R. Environmental and Economic Benefits from the Phase-out of Residential Oil Heating: A Study from the Aosta Valley Region (Italy). *Sustainability* **2019**, *11*, 3633. [CrossRef]
24. Sternberg, A.; Bardow, A. Power-to-What?—Environmental Assessment of Energy Storage Systems. *Energy Environ. Sci.* **2015**, *8*, 389–400.
25. Casasso, A.; Tosco, T.; Bianco, C.; Bucci, A.; Sethi, R. How Can We Make Pump and Treat Systems More Energetically Sustainable? *Water* **2019**, *12*, 67. [CrossRef]

26. Filippini, M.; Hunt, L.C.; Zorić, J. Impact of Energy Policy Instruments on the Estimated Level of Underlying Energy Efficiency in the EU Residential Sector. *Energy Policy* **2014**, *69*, 73–81. [CrossRef]
27. Koengkan, M.; Fuinhas, J.A.; Osmani, F.; Kazemzadeh, E.; Auza, A.; Alavijeh, N.K.; Teixeira, M. Do Financial and Fiscal Incentive Policies Increase the Energy Efficiency Ratings in Residential Properties? A Piece of Empirical Evidence from Portugal. *Energy* **2022**, *241*, 122895. [CrossRef]
28. Fuinhas, J.A.; Koengkan, M.; Silva, N.; Kazemzadeh, E.; Auza, A.; Santiago, R.; Teixeira, M.; Osmani, F. The Impact of Energy Policies on the Energy Efficiency Performance of Residential Properties in Portugal. *Energies* **2022**, *15*, 802. [CrossRef]
29. Ballarini, I.; Pichierri, S.; Corrado, V. Tracking the Energy Refurbishment Processes in Residential Building Stocks. The Pilot Case of Piedmont Region. *Energy Procedia* **2015**, *78*, 1051–1056. [CrossRef]
30. TABULA. EPISCOPE Joint EPISCOPE and TABULA Website. Available online: https://episcope.eu/welcome/ (accessed on 3 November 2021).
31. Corrado, V.; Ballarini, I.; Corgnati, S.P. National Scientific Report on the TABULA Activities in Italy. Available online: http://bit.ly/TABULA_IT (accessed on 10 January 2020).
32. Ballarini, I.; Corgnati, S.P.; Corrado, V. Use of Reference Buildings to Assess the Energy Saving Potentials of the Residential Building Stock: The Experience of TABULA Project. *Energy Policy* **2014**, *68*, 273–284. [CrossRef]
33. MISE Requisiti Tecnici per l'accesso Alle Detrazioni Fiscali Per La Riqualificazione Energetica Degli Edifici—Cd. Ecobonus. (Technical Requirements for the Eligibility to Tax Deductions for Building Energy Refurbishment). Available online: https://bit.ly/DM20200806 (accessed on 3 November 2021).
34. Repubblica Italiana DPR 412/1993—Regolamento Recante Norme Per La Progettazione, l'installazione, l'esercizio e La Manutenzione Degli Impianti Termici Degli Edifici Ai Fini Del Contenimento Dei Consumi Di Energia, in Attuazione Dell'art. 4, Comma 4, Della Legge 9 Gennaio 1991, n. 10. (Regulation for Designing, Installation, Operation and Maintenance of HVAC Systems). Available online: https://bit.ly/DPR41293 (accessed on 3 November 2021).
35. TEP Software ANIT. Available online: https://www.anit.it/software-anit/ (accessed on 29 March 2021).
36. UNI UNI/TS 11300 Standard Series. Available online: https://www.cti2000.eu/la-uni-ts-11300/ (accessed on 24 January 2022).
37. ISO. ISO 13790:2008. Energy Performance of Buildings—Calculation of Energy Use for Space Heating and Cooling. Available online: https://www.iso.org/standard/41974.html (accessed on 16 March 2022).
38. UNI UNI 10349:2016 Standard Series. Available online: https://bit.ly/3HSwyoD (accessed on 8 March 2022).
39. ISTAT. I Consumi Energetici Delle Famiglie (Energetic Consumption of Italian Families, Divided by Region). Available online: https://www.istat.it/it/archivio/142173 (accessed on 17 September 2018).
40. BLOCON EED—Earth Energy Designer. Version 4.20. Available online: https://buildingphysics.com/eed-2/ (accessed on 8 July 2021).
41. Eskilson, P. Thermal Analysis of Heat Extraction Boreholes. Available online: http://bit.ly/2JCl6Be (accessed on 8 July 2021).
42. Dalla Santa, G.; Galgaro, A.; Sassi, R.; Cultrera, M.; Scotton, P.; Mueller, J.; Bertermann, D.; Mendrinos, D.; Pasquali, R.; Perego, R.; et al. An Updated Ground Thermal Properties Database for GSHP Applications. *Geothermics* **2020**, *85*, 101758. [CrossRef]
43. Bartolini, N.; Casasso, A.; Bianco, C.; Sethi, R. Environmental and Economic Impact of the Antifreeze Agents in Geothermal Heat Exchangers. *Energies* **2020**, *13*, 5653. [CrossRef]
44. Repubblica Italiana D.Lgs. 28/2011—Attuazione Della Direttiva 2009/28/CE Sulla Promozione Dell'uso Dell'energia Da Fonti Rinnovabili, (Legislative Decree 28/2011 on the Promotion of Renewable Energy Sources, Applying the EC Directive 2009/28/EC). Available online: https://bit.ly/DLgs28_2011 (accessed on 4 January 2019).
45. FERROLI. Price List. Available online: https://bit.ly/3Il4KJQ (accessed on 15 July 2021).
46. FONDERIE SIME. Price List. Available online: https://bit.ly/3qfHTZT (accessed on 16 July 2021).
47. MESCOLI. Price List. Available online: https://bit.ly/3we63ru (accessed on 3 September 2021).
48. BOSCH. Price List. Available online: https://bit.ly/3wg8N7N (accessed on 2 September 2021).
49. VAILLANT. Price List. Available online: https://bit.ly/3KRkoy5 (accessed on 10 September 2021).
50. VIESSMANN. Price List. Available online: https://bit.ly/3KXDnr3 (accessed on 30 August 2021).
51. Doseva, N.; Chakyrova, D. Life Cycle Cost Analysis of Different Residential Heat Pump Systems. *E3S Web Conf.* **2020**, *207*, 01014. [CrossRef]
52. Vering, C.; Maier, L.; Breuer, K.; Krützfeldt, H.; Streblow, R.; Müller, D. Evaluating Heat Pump System Design Methods towards a Sustainable Heat Supply in Residential Buildings. *Appl. Energy* **2022**, *308*, 118204. [CrossRef]
53. Regione Piemonte Prezzario delle Opere Pubbliche della Regione Piemonte (Price List of Public Works in the Piedmont Region). Available online: https://bit.ly/3o5aDCI (accessed on 31 March 2021).
54. AERMEC. Italian Price List. Available online: https://bit.ly/3COeVWb (accessed on 1 September 2021).
55. DAIKIN. Price List. Available online: https://bit.ly/3JrK7Ny (accessed on 4 September 2021).
56. Blum, P.; Campillo, G.; Kölbel, T. Techno-Economic and Spatial Analysis of Vertical Ground Source Heat Pump Systems in Germany. *Energy* **2011**, *36*, 3002–3011. [CrossRef]
57. BRGM. Action d'accompagnement Pour Le Développement de La Chaleur Géothermale. Convention ADEME-BRGM 2011—Synthèse. Rapport Final [Action for the Development of Geothermal Heat. Final Report]. Available online: http://infoterre.brgm.fr/rapports/RP-60843-FR.pdf (accessed on 25 February 2021).

58. CCIAA. Torino Turin Chamber of Commerce—Biweekly Update on Fuel Prices. Available online: https://www.to.camcom.it/listino-quindicinale (accessed on 25 February 2021).
59. GME. Italian Energy Market Manager (GME) Website—Sale Market Price. Available online: https://www.mercatoelettrico.org/it/ (accessed on 25 February 2021).
60. CLIVET. CLIVET Manufacturer Website. Available online: https://www.clivet.com/ (accessed on 7 May 2021).
61. GSE Italian Energy Service Manager (GSE)—Explanation of the on-Site Exchange Mechanism. Available online: https://www.gse.it/servizi-per-te/fotovoltaico/scambio-sul-posto (accessed on 5 May 2021).
62. Hansen, K. Decision-Making Based on Energy Costs: Comparing Levelized Cost of Energy and Energy System Costs. *Energy Strategy Rev.* **2019**, *24*, 68–82. [CrossRef]
63. Eicher, S.; Hildbrand, C.; Kleijer, A.; Bony, J.; Bunea, M.; Citherlet, S. Life Cycle Impact Assessment of a Solar Assisted Heat Pump for Domestic Hot Water Production and Space Heating. *Energy Procedia* **2014**, *48*, 813–818. [CrossRef]
64. Latorre-Biel, J.-I.; Jimémez, E.; García, J.L.; Martínez, E.; Jiménez, E.; Blanco, J. Replacement of Electric Resistive Space Heating by an Air-Source Heat Pump in a Residential Application. Environmental Amortization. *Build. Environ.* **2018**, *141*, 193–205. [CrossRef]
65. Pike, C.; Whitney, E. Heat Pump Technology: An Alaska Case Study. *J. Renew. Sustain. Energy* **2017**, *9*, 061706. [CrossRef]
66. Ma, Z.J.; Thomas, S. Reliability and Maintainability in Photovoltaic Inverter Design. In Proceedings of the 2011 Annual Reliability and Maintainability Symposium, Lake Buena Vista, FL, USA, 24–27 January 2011; pp. 1–5. [CrossRef]
67. Abdul-Ganiyu, S.; Quansah, D.A.; Ramde, E.W.; Seidu, R.; Adaramola, M.S. Techno-Economic Analysis of Solar Photovoltaic (PV) and Solar Photovoltaic Thermal (PVT) Systems Using Exergy Analysis. *Sustain. Energy Technol. Assess.* **2021**, *47*, 101520. [CrossRef]
68. Nitkiewicz, A.; Sekret, R. Comparison of LCA Results of Low Temperature Heat Plant Using Electric Heat Pump, Absorption Heat Pump and Gas-Fired Boiler. *Energy Convers. Manag.* **2014**, *87*, 647–652. [CrossRef]
69. Casasso, A.; Puleo, M.; Panepinto, D.; Zanetti, M. Economic Viability and Greenhouse Gas (GHG) Budget of the Biomethane Retrofit of Manure-Operated Biogas Plants: A Case Study from Piedmont, Italy. *Sustainability* **2021**, *13*, 7979. [CrossRef]
70. Mohammadpourkarbasi, H.; Sharples, S. Appraising the Life Cycle Costs of Heating Alternatives for an Affordable Low Carbon Retirement Development. *Sustain. Energy Technol. Assess.* **2022**, *49*, 101693. [CrossRef]
71. Poponi, D.; Basosi, R.; Kurdgelashvili, L. Subsidisation Cost Analysis of Renewable Energy Deployment: A Case Study on the Italian Feed-in Tariff Programme for Photovoltaics. *Energy Policy* **2021**, *154*, 112297. [CrossRef]
72. Reber, T.J.; Beckers, K.F.; Tester, J.W. The Transformative Potential of Geothermal Heating in the U.S. Energy Market: A Regional Study of New York and Pennsylvania. *Energy Policy* **2014**, *70*, 30–44. [CrossRef]
73. AFPG. Etude Techno-Économique de La Géothermie de Surface (Techno-Economic Study of Shallow Geothermal Energy). Available online: https://bit.ly/3G80L2G (accessed on 1 December 2021).
74. Luo, J.; Zhang, Y.; Rohn, J. Analysis of Thermal Performance and Drilling Costs of Borehole Heat Exchanger (BHE) in a River Deposited Area. *Renew. Energy* **2020**, *151*, 392–402. [CrossRef]
75. IEA. Cumulative Capacity and Capital Cost Learning Curve for Vapour Compression Applications in the Sustainable Development Scenario, 2019–2070. Available online: https://bit.ly/3D86rrf (accessed on 9 September 2021).
76. Köhler, B.; Stobbe, M.; Garzia, F. EU H2020 Project CRAVEzero. Deliverable D.4.1. Guideline II: NZEB Technologies: Report on Cost Reduction Potentials for Technical NZEB Solution Sets. Available online: https://bit.ly/3G0BUi2 (accessed on 9 September 2021).
77. European Commission, Joint Research Centre. *Li-Ion Batteries for Mobility and Stationary Storage Applications: Scenarios for Costs and Market Growth*; Publications Office: Luxembourg, 2018.
78. IEA. What Is Behind Soaring Energy Prices and What Happens Next? Available online: https://bit.ly/3ie70I5 (accessed on 16 March 2022).
79. Trading Economics EU Natural Gas—2022 Data—2010–2021 Historical. Available online: https://bit.ly/3ibcKm7 (accessed on 16 March 2022).
80. Martinopoulos, G.; Papakostas, K.T.; Papadopoulos, A.M. A Comparative Review of Heating Systems in EU Countries, Based on Efficiency and Fuel Cost. *Renew. Sustain. Energy Rev.* **2018**, *90*, 687–699. [CrossRef]
81. Novelli, A.; D'Alonzo, V.; Pezzutto, S.; Poggio, R.A.E.; Casasso, A.; Zambelli, P. A Spatially-Explicit Economic and Financial Assessment of Closed-Loop Ground-Source Geothermal Heat Pumps: A Case Study for the Residential Buildings of Valle d'Aosta Region. *Sustainability* **2021**, *13*, 12516. [CrossRef]
82. Bertelsen, N.; Vad Mathiesen, B. EU-28 Residential Heat Supply and Consumption: Historical Development and Status. *Energies* **2020**, *13*, 1894. [CrossRef]
83. POLIMI. Smart Building Report 2020. Energy Efficiency and Digital Technologies to Improve the Building Sector (Original Title in Italian). Available online: https://bit.ly/35Xr3YT (accessed on 14 March 2022).
84. Barnes, J.; Bhagavathy, S.M. The Economics of Heat Pumps and the (Un)Intended Consequences of Government Policy. *Energy Policy* **2020**, *138*, 111198. [CrossRef]
85. Boldrin, G. Transfer of Credit and Discount on Invoice for Tax Deductions: Analysis and Evolution (Original Title in Italian). Master's Thesis, University of Venezia "Ca' Foscari", Venezia, Italy. Available online: https://bit.ly/3CGuD5o (accessed on 14 March 2022).

86. EHPA. European Heat Pump Market and Statistic Report 2020 & Stats Tool. Available online: http://www.stats.ehpa.org/hp_sales/country_cards/ (accessed on 15 March 2022).
87. Eurostat. Final Energy Consumption in Households by Type of Fuel. Available online: https://ec.europa.eu/eurostat/databrowser/view/ten00125/default/table?lang=en (accessed on 15 March 2022).
88. Bioenergy Europe Bioenergy Europe Statistical Report. 2020. Available online: https://bit.ly/36kZPuR (accessed on 15 March 2022).
89. Sarigiannis, D.A.; Karakitsios, S.P.; Kermenidou, M.V. Health Impact and Monetary Cost of Exposure to Particulate Matter Emitted from Biomass Burning in Large Cities. *Sci. Total Environ.* **2015**, *524–525*, 319–330. [CrossRef]
90. Verkerk, P.J.; Fitzgerald, J.B.; Datta, P.; Dees, M.; Hengeveld, G.M.; Lindner, M.; Zudin, S. Spatial Distribution of the Potential Forest Biomass Availability in Europe. *For. Ecosyst.* **2019**, *6*, 5. [CrossRef]
91. Malico, I.; Nepomuceno Pereira, R.; Gonçalves, A.C.; Sousa, A.M.O. Current Status and Future Perspectives for Energy Production from Solid Biomass in the European Industry. *Renew. Sustain. Energy Rev.* **2019**, *112*, 960–977. [CrossRef]
92. Ruiz, P.; Nijs, W.; Tarvydas, D.; Sgobbi, A.; Zucker, A.; Pilli, R.; Jonsson, R.; Camia, A.; Thiel, C.; Hoyer-Klick, C.; et al. ENSPRESO—An Open, EU-28 Wide, Transparent and Coherent Database of Wind, Solar and Biomass Energy Potentials. *Energy Strategy Rev.* **2019**, *26*, 100379. [CrossRef]

energies

MDPI

Article

Renovation of Modernist Housing Developments in the Pursuit of Modernity for Well-Being and Clean Energy

Adam Turecki [1], Marcin Tur [1], Bartosz Czarnecki [1,*], Krystyna Januszkiewicz [2] and Piotr Fiuk [2]

1 Faculty of Architecture, Bialystok University of Technology, 15-893 Bialystok, Poland; a.turecki@pb.edu.pl (A.T.); m.tur@pb.edu.pl (M.T.)
2 Faculty of Architecture, West Pomeranian University of Technology in Szczecin, 70-310 Szczecin, Poland; krystyna.januszkiewicz@zut.edu.pl (K.J.); piotr.fiuk@zut.edu.pl (P.F.)
* Correspondence: b.czarnecki@pb.edu.pl

Abstract: The research concerns renovation variants for modernist housing estates built in Szczecin, Poland during 1918–1925 and in Bialystok, Poland during 1950–1990. These buildings are now substandard in many ways; functionally, aesthetically, technically, and ecologically they do not fulfil the current energy efficiency standards. Some of them have architectural heritage, so not all energy-saving technologies can be applied. Renovations must include energy-saving improvements and the use of renewable energy sources. Equally important is the well-being of residents, meaning the quality of the apartments should be increased. The aim of this research was to analyze the renovation options in terms of energy efficiency and well-being criteria, as well as in relation to the cultural value of the buildings. The simplified energy calculation method was used to check the present buildings' energy demands to compare them with retrofitting results. Three retrofitting possibilities were considered: low-cost, current standards, and near-zero energy. The results show that without EU financial aid, which will soon be introduced under the "Renovation Wave" program, such modernization projects will be difficult, making the target of 55% CO_2 emission reductions compared to 1990 levels by 2030 impossible.

Keywords: architecture; "Renovation Wave"; energy-efficiency; CO_2 emissions; well-being

Citation: Turecki, A.; Tur, M.; Czarnecki, B.; Januszkiewicz, K.; Fiuk, P. Renovation of Modernist Housing Developments in the Pursuit of Modernity for Well-Being and Clean Energy. *Energies* **2022**, *15*, 3737. https://doi.org/10.3390/en15103737

Academic Editors: Cynthia Hou and Joseph H.K. Lai

Received: 19 April 2022
Accepted: 14 May 2022
Published: 19 May 2022

1. Introduction

Achieving the EU's 'Fit For 55' target by 2030 will be impossible without renovating buildings built in the previous century. Industrial development at that time accelerated the growth of many cities. New districts began to rise around existing city centers. They were shaped according to the urban and architectural ideas of the modernist movement, appeared in early 20th century. Today, these buildings are the majority in many cities. About 35% of them are over 50 years old [1]. The issues of energy efficiency and CO_2 emissions were marginal until the end of the 20th century [2]. Now we know that buildings consume about 40% of the total energy produced worldwide and that the associated CO_2 emissions have a significant impact on climate change [3]. Studies on the use of solar energy were maybe the most spectacular and easy to explain for publicity, but in the climate zone of central Europe the use of technologies based on geothermal energy is more efficient.

Research on the energy efficiency of buildings began in the last century. In 1939, the Massachusetts Institute of Technology (MIT) constructed "Solar 1" [4] (architect: Vannevar Bush)—probably the world's first research building—to test the possibility of heating using solar energy absorbed by roof collectors and stored in a large insulated tank in the basement. In the summer, the ability to passively cool the facility was checked. By 1978, MIT had built another five research houses. "Zero-Energy House", seen as one of the first nearly zero-energy homes in the world, was built in 1974 at the Technical University of Denmark in Copenhagen (architects: Vagn Korsgaard, Torben Esbensen) [5]. In 1992, the Fraunhofer Institute for Solar Energy Systems (ISE) in Freiburg in Germany completed "Das

Energieautarke Solarhaus"—Autonomous Solar House (architects: Hölken and Berghoff). The building's energy autonomy was achieved through the use of multiple technologies and through storing the hydrogen produced during the electrolysis of water using electricity generated by a photovoltaic system from solar radiation, a pioneering feat in architecture. The hydrogen was then converted into electricity in fuel cells during periods of solar energy shortage [6].

These buildings were the highest technical achievements of their time. They inspired further research that resulted in increasingly ambitious energy targets. In recent decades, the standards of energy consumption requirements for buildings have been steadily raised (energy efficient <70 $kWh/m^2{\cdot}y$, low-energy <40 $Wh/m^2{\cdot}y$, passive <15 $Wh/m^2{\cdot}y$, zero energy = 0 $Wh/m^2{\cdot}y$, +energy–building–producer supplying surplus energy to the electrical grid). The last decades show that such buildings are structurally durable and may serve for many years. The problems are their technical equipment and energy efficiency.

1.1. New Criteria for Energy Efficiency

Recent research by climatologists published in the Reports of the Intergovernmental Panel on Climate Change (ICPP) has changed the criteria for assessing buildings. In addition to energy consumption requirements, new ones have been introduced, emphasizing CO_2 emissions associated with the construction and operation of buildings. In 2005, American architect Edward Mazria introduced a new target—"Architecture 2030"; that is, zero-carbon buildings by 2030 [7].

The above criteria were introduced into many strategies and programs, both global, such as the "Paris Agreement" in 2015, and local, which in the EU included the "European Green Deal" in 2019, "Fit For 55" in 2021, and the current initiative "100 Climate-Neutral and Smart Cities by 2030" from 2021. Unfortunately, the present criteria for the building regulations set lower requirements. In Poland in 2021, the Building Regulations for multifamily buildings allow primary energy consumption rates for heating and cooling of buildings of up to 65 $kWh/(m^2\text{-}yr)$. This differs significantly from the zero-energy building standards, which the EU already recommended in 2020. They apply only to newly constructed buildings—i.e., a small proportion of those existing today.

Therefore, the EU program "Renovation Wave" [8], launched in 2020, is very important. It indicates that only 1% of old buildings are now renovated yearly and that the rate of renovation should be increased to 4% per year. A 1% rate would extend the renovation process to the end of the 21st century, meaning a minimum of 4% will allow the work to be completed in 25 years and achieve the climate neutrality goals for cities by 2050.

1.2. Energy Efficiency in the Context of Well-Being

Improving existing buildings is particularly important in Poland—this is confirmed by the listing of 36 Polish cities on the register of 50 European cities with the highest smog levels in 2018 [8] and the frequent negative image of air pollution in Poland visible on web maps of smog measurements in Europe (Figure 1) [9].

Although they are mainly based on PM10 and PM2.5 air dust measurements, the presence of particulate matter is mainly the result of burning coal and wood—practices that are strongly associated with CO_2 emissions. Figure 1 shows the data under specific conditions—namely in winter, with an outside temperature lower than −5 °C, during an almost windless evening, without precipitation, and with inversion. Such weather conditions are rare but this is when the effects of burning fossil fuels are clearly visible. Under different weather conditions, for example with stronger wind, the measured dustiness in the air is lower but at the same temperature (=−5 °C) more energy is needed to heat the buildings and the real CO_2 emissions level is higher. Pollen meters do not show this, but the amount of greenhouse gases in the atmosphere increases. The same happens with high levels of emissions from CHP plants. Therefore, even urban housing estates connected to city heating networks contribute to harmful climate change.

Figure 1. Image showing high levels of air pollution in Poland visible on the Airly web map of smog measurements in Europe on 24 January 2022, adapted from Ref. https://airly.org/map/pl/; accessed on 24 January 2022, 00:24

Energy efficiency is associated with good quality of life. As a criterion, a good indicator is a well-being approach. In its popular meaning, it is a sense of comfort, although well-being theory is much deeper itself. Martin Seligman tried to describe key elements influencing human self-sense, resulting in PERMA^{TM}: Positive Emotions, Engagement, Relationships, Meaning, and Achievement and Accomplishment [9]. This concept seems to be only of the mental sphere, but we should understand that this sphere is strongly connected to physical conditions of life. Only if a human being's basic needs are assured (see the Maslow categories of human needs pyramid—physiologic needs, safety, etc.) with their higher needs be assured, i.e., emotional needs. Therefore, the improvement of energy efficiency cannot be achieved through the worsening of physical accommodation conditions, nor via extreme increases in life costs. What is the scale of renovation needs, and are housing communities able to finance them themselves?

Apart from issues related to a lack of energy efficiency and the associated high charges for energy carriers, which have increased rents, especially recently, beyond the affordability of poorer residents, postwar multifamily buildings have a number of other weaknesses. The flats are small and were regarded as too small in the years they were built. However, at that time their prices were much lower than their building costs. State cooperatives took over private building sites for negligible compensation. No one counted the true cost of development. Housing was a kind of social assistance. Under such conditions, residents always expected larger flats. Today, when flats are sold for the market price, which is very high in relation to wages, expectations of large flats have diminished and the existing old ones are larger than what new developers often offer today. Monotonous interior design, including the use of primitive finishing materials, ugly furniture, and bad colours, usually arranged by the owners, were the typical aspects of living in such buildings in these times. Today, specialized companies providing good design details and better materials renovate these flats, which means their appearance meets high aesthetic standards. The next problem with these buildings is the external elements, namely the balconies and loggias. Many small flats do not have them at all, while in larger ones they are so small that sometimes it is impossible to sit on a chair on them. When a flat is small, a balcony or loggia is an important element, as it compensates for the cramped living space. A cardinal disadvantage of these blocks is the staircases, which are narrow and primitively finished in bad colours, which additionally provokes destructive behaviour, including vandalism and graffiti. Some of the five-storey blocks were built without lifts. The people that started living in them were usually young and climbing a few floors was not so difficult. Today, most of the people living in these blocks are elderly people who can find it difficult to walk even on flat ground—flats on floors higher than the ground floor have become prisons for them. Additionally, access is an issue for mothers with children riding in prams. Some of the flats on the ground floor have small gardens. These would be great for disabled people or families with small children, but to get to the gardens one has to climb several steps because the ground floor level is usually raised more than one meter above the ground. The disadvantages mentioned above mean that anyone with a little more money is trying to "escape" from the blocks of flats to detached houses, usually in the distant suburbs.

The social structure of the blocks of flats is changing. There is a predominance of elderly residents, who often have trouble paying their bills. Large flats are often inhabited by a single person living on a small pension. Turning off the heating in mostly empty rooms results in the beginning of degradation of the building materials. The vacant flats are occupied by poorer people or temporary residents, who often do not care about the technical condition of the flats. The blocks are beginning to be seen as poor places to live. Youth are moving to the suburbs. City centers are starting to decline. This has also resulted in increased traffic congestion, morning and evening traffic jams, and additional emissions. The operational costs of cities are rising, and a lack of resources for vital social functions and necessary investment has arisen. This crisis has forced active youth and businesses to migrate to other towns. These cities are entering a period of depopulation and gentrification, the beginning of a process of decline; we have seen such a process previously in Detroit.

Postwar buildings also have their advantages in many cities, as they are located in downtowns or in their close vicinity. They have large courtyards with decades-old trees. The technical condition of the buildings ensures that they will function for many years, although it is necessary to renovate them. A successful example is the renovation of a complex of three fifteen-storey modernist blocks in the Grand Parc neighbourhood in Bordeaux. Anne Lacaton and Jean-Philippe Vassal are the creators of modernization. The motto of their projects is: "Never demolish, never remove or replace, always add, transform, and reuse!" [10]. A framework was added to the existing walls to create large outdoor loggias with sliding glass walls, creating additional outdoor "green rooms". The existing small windows were replaced with larger ones reaching to the floor. This brightened up the interiors, also improving the views from the windows, which gave the flats a more contemporary look. Additional glass lifts were built next to the staircases. The renovation was carried out without displacing residents. The comfort, aesthetics, use value, and energy efficiency of the flats have increased. Energy consumption has decreased thanks to the glass buffers, and probably the rents. The effect of glazing loggias to reduce energy losses in dwellings was described in 2011 by one of the authors [11].

The project was recognized by the Pritzker Prize awarded to the authors in 2021, and we should continue with the methods, solutions, and logistics of the redevelopment they introduced (Figure 2). The Lacaton and Vassal project includes the same kinds of buildings as most of those reported in our article, namely 1960s–1980s multifamily houses made using industrialized building technologies. This project shows that it is possible to incorporate both formal and utility values into such buildings.

Wider research including human factors in such approaches to the modernization of multifamily houses in Italy was described by Lucchi and Delera [12].

Figure 2. Renovation of a complex of three fifteen-storey blocks in the Grand Parc in Bordeaux. Pritzker Prize 2021. Architects: Anne Lacaton and Jean-Philippe Vassal, adapted from Ref. https://www.pritzkerprize.com/sites/default/files/inline-files/Transformation%20of%20G%2C%20H%2C%20I%20Buildings%201.jpg photo Philippe Ruault, (accessed on 13 May 2022).

1.3. Study Objectives

In Europe after World War II, the acute problem was the lack of housing. In response to this, industrialized housing technologies were developed in both Western and Eastern European countries. Such properties are inhabited by a very large part of the population. Usually, their energy efficiency in terms of the thermal insulation is unsatisfactory. This, in combination with their very significant share of the housing stock, causes their large impact on energy consumption in the global context, as well as their contribution to the greenhouse effect. At the same time, the generally significant share of people from economically weaker groups among the residents means that expecting a greater participation in the energy transformation costs from the residents of such properties raises serious doubts.

The aim of this research, in general, was to determine the needs in terms of achieving the currently expected energy efficiency targets for these housing resources and finally to answer the question of whether it is possible to finance these works from the residents' own resources or whether it requires external support.

1.3.1. Objective 1

The first aim of the research was to estimate the differences between the actual and expected energy standards based on selected examples of buildings from different periods and of different types, and then to determine the energy demands.

1.3.2. Objective 2

The level of CO_2 emissions connected with energy consumption, in particular with modernization efforts, was checked. A comparison was made of CO_2 emission reductions achieved through renovations with the targets set by the EU in the "Fit For 55" program.

1.3.3. Objective 3

An analysis of the results and basic financial calculations of the costs of the various renovation variants, by converting the results into sums per building inhabitant, was performed with reference to the level of income of the population to answer the question of whether such modernization is possible with the residents' own resources and to what extent it is necessary to support this with external funds.

2. Materials and Methods

According to the aims of the research, the required data including the parameters of buildings in the housing estates. The authors decided to compare the energy efficiency standards of buildings from different periods. Therefore, the selected estates seemed to be representative of completely different periods.

2.1. Modernist Housing Estates in the 1920s and 1930s in Szczecin: The Solution to Social, Infrastructural, and Technological Problems

In the 1920s and 1930s in Szczecin [13], there were attempts to develop new solutions for urban layouts, continuing the process of the modernization of excessively dense European industrial cities [14]. The new urban solutions combined with the rational planning of apartments reflected the ideas of the modernist movement [15] in architecture, popularized by the international community of designers and avant garde artists [16]. The structure, based on the "residential units" and "housing estates" extracted from the spaces of traditional cities, determined the perspectives for the rational development of urban layouts, resulting in the use of modern (at the beginning of the 20th century) infrastructural and technological solutions providing adequate housing conditions for the less wealthy members of society [17].

The historical rule of peripheral building developments in old cities, involving large quarters and courtyards with outbuildings, was replaced by rectangular buildings or the free planning of new homes [18]. The locations of multifamily buildings were adapted to the configuration of the terrain, creating housing estates with the organic planning of streets and buildings. Where there were differences in height, freestanding buildings with a semicircular or oval outline were created. Green areas isolated from traffic were prepared as recreational facilities for residents of the housing estates.

To facilitate pedestrian traffic inside multifamily developments, which were erected using the modest funds available to housing cooperatives and building societies, external traffic galleries were constructed, acting equivalently to balconies in bigger apartments but with higher standards. The internal traffic within modest multifamily developments was solved through the construction of corridors (with several or over a dozen apartments per floor) and separate staircases (with access to between 2 to 4 apartments on subsequent levels). In the constructions of multifamily buildings, designers used traditional masonry (bricks with air gaps or slag concrete blocks), as well as new technologies such as reinforced concrete and steel. The staircases were made in a number of ways—using traditional wood, with wooden steps on a reinforced concrete or ceramic slab, and using reinforced concrete only. The ceilings over basements were built with ceramics or reinforced concrete, while wood was usually used on the levels with apartments.

Regarding aesthetics, the appearance of the facades from the street and the interior of the estate was unified. The plastered walls were enriched with ceramic decorative elements such as plinths, window and door trims, inter-window strips horizontally crowning the shapes, as well as repetitive segments of staircases and external traffic galleries. The overhanging balcony and gallery elements contained plastic accents of lights and shadows on homogeneous surfaces of cubist blocks. The geometric metalwork elements, the corner glazing on the verandas, and the window arrangements in staircases constituted additional accents in the restrained compositions (Figure 3).

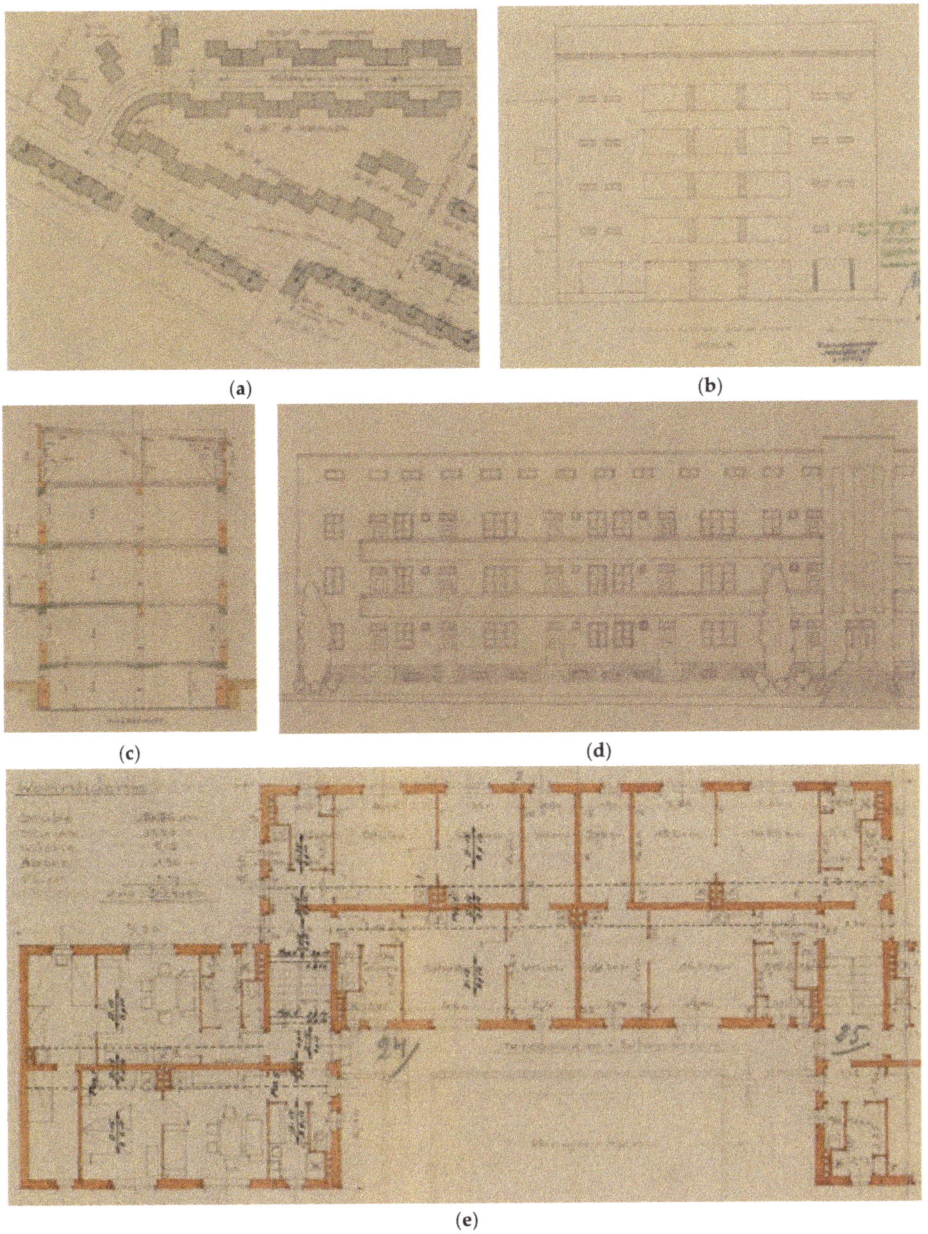

(a) (b) (c) (d) (e)

Figure 3. Szczecin (**a**) urban plan of a modernist housing estate between Boelcke Strasse, Richthofen Strasse, and Reichswehr Strasse (Stanisława Brzozowskiego St., Karola Huberta Rostworowskiego St., Klemensa Janickiego St.) from 1931. (**b**) Front elevation of a multiapartment building with loggias in Boelcke Strasse (Stanisława Brzozowskiego St.) from 1931. (**c**,**d**) Cross-sections and elevation of a building with external galleries in Reichswehr Strasse (Klemensa Janickiego St.) from 1932. (**e**) Multifamily building with differentiated building lines in Richthofen Strasse (Karola Huberta Rostworowskiego St.), showing ground floor plan from 1931 from the State Archive in Szczecin, Construction Supervision Records, signature 10806, reprinted from Ref. Szczecin Municipal Archives.

To ensure the economic use of funds and to facilitate the construction process, the designers used repetitive residential sections. Kitchen, bathroom, and furniture elements were also standardized, the creation of which included features characteristic of the preliminary phase of industrial production. The formation of rational layouts of multifamily suburban housing estates was based on the system of industrial production (functional homes were erected in the 1930s in Berlin based on model designs with the use of prefabricated construction elements) [19]. Economic solutions were devised to satisfy the needs of the increasing numbers of residents in industrialized urban hubs. There was a need for solutions that would be within reach of families with average salaries [20]. The new multifamily buildings constructed in the 1920s and 1930s in the suburban housing estates of Szczecin fulfilled the necessary functional and utilitarian requirements, containing new apartments with multiple rooms, as well as more modest social flats [21].

The designs for Szczecin's housing estates from the 1920s and 1930s were based on the solutions developed in Berlin [22] and the urban architectural modernization concepts implemented after World War I in Frankfurt am Main [23]. From the end of the 18th century and beginning of the 19th century, Szczecin had been economically connected to Berlin in terms of the use of the transshipment facilities in the port by the Oder. In the 19th century, there were many analogies in terms of construction investment, with references to Berlin's typology of tenement houses and the urban plans from the industrial era developed in the capital (Figure 4).

2.2. *Postwar Housing Estates in Bialystok*

During WWII, Bialystok was one of the most destroyed cities in Poland. About 80% of the city center was destroyed during the four passages of the German and Soviet armies in 1939–1944 (Figure 5). The few partially preserved buildings were restored and the demolished ones were replaced with new ones, but the number of flats was much lower, along with a lower number of rooms in each flat and smaller internal surface area. They were reconstructed in accordance with the historical layout of the streets and quarters, with significantly enlarged backyards. They were still topped with pitched roofs, sometimes with traditional facade decorations. The enormity of the destruction allowed for the correction of the street grid, resulting in chaotic spaces connecting the new streets and the preserved buildings, built along the prewar streets. In the following years, after Stalin's death, a political change took place, which resulted in the abandonment of historicism and a return to modernist architecture. New districts appeared outside the historical center. Until 1969, buildings were still constructed in a traditional way, with brick used as the basic material. However, the new buildings were characterized by substandard living conditions.

The growth of the city associated with postwar industrialization and the arrival of large numbers of people to Bialystok was associated with a major housing shortage problem. In response to this, both in Western countries and behind the Iron Curtain, industrialized housing technologies were developed. Housing estates consisting of such buildings were built in Bialystok from 1969 to the mid-1990s. As a result, nowadays a very large proportion of the population lives in such buildings. In Poland, the share of inhabitants of buildings erected using various technologies of this type was estimated in the year 1999 as ca. 10–12 million people [24]. Such buildings have not been demolished; changes to the demographics could have resulted only from decreasing the average number of family members and from increasing the number of single people, so the current population living in such buildings in Poland is likely 8 million inhabitants, and maybe a little more.

In Poland, there are ca. 4 million flats in prefabricated buildings. According to various data sources, 10 to 12 million Poles inhabit these flats. In most of the main cities, large-panel buildings account for more than 30% of the total housing stock [25].

This constitutes around 30% of the total population of Poland. These buildings were erected in completely different economic contexts, both from energy efficiency and ecological perspectives. Their energy efficiency in terms of the thermal insulation usually leaves much to be desired. Combined with their very significant share of the housing stock (Figure 6), this results in their high impact on energy consumption globally, as well as their significant contribution to the greenhouse effect. At the same time, the generally significant share of people from economically weaker groups among the inhabitants means that expecting the inhabitants of such buildings to make significant contributions to the energy transformation costs raises serious doubts.

Figure 4. Szczecin residential buildings with outdoor galleries in Reichswehr Strasse (Klemensa Janickiego), source: authors—photo: K. Januszkiewicz, P. Fiuk.

Figure 5. Bialystok city center in 1944: ca. 80% of the buildings were destroyed during WWII, reprinted from Ref. The Library of USA Congress.

Figure 6. Bialystok city center in 2021. The locations of multifamily buildings are marked in color, adapted from Ref. Municipality of Bialystok, GIS Department.

2.3. Prefabricated Buildings

The increase in the urban population convinced the country's authorities to use prefabricated construction systems (in England they were called "large-panel system buildings" (LPS), in Germany "Plattenbau"). A number of such technologies have been tested in Poland, including W-70, WK-70, S, WUF-T, WWP, and OW1700 systems. In 1967, one of the more popular systems in Poland was developed—OW-T/67 (this is an acronym for economical, large-scale, and typical). The ground floor plan was usually rectangular; two heights were mainly used—5-storey (without lift) and 11-storey. A section of each building had a central staircase, usually with two or three flats on one floor. Occasionally external galleries were used. The smallest of the created buildings usually had three sections measuring 40–50 m in length, with the longest measuring Poland 860 m. The OW-T/67 system provided flats of various sizes (the number indicates the number of residents): M1—20–23 m^2; M2—30–32 m^2; M3—38–45 m^2; M4—48–54 m^2; M5—57–60 m^2; M6—62–66 m^2; M7—72–75 m^2. Usually, there were three flats of different sizes on one floor (M3 + M1 + M4, M3 + M2 + M5,etc.), although sometimes only two (M4 + M5−M7). Each flat had a bathroom, kitchen, and a different number of rooms, while the larger flats (M3−M7) also had a balcony or loggia.

In 1968, in Bialystok in the suburban fields near the Bojary district, a factory was built to produce building elements used in the OW-T/67 system. The new district of Piasta was planned around it. In 1969, at 3 Piastowska Street, the first prefabricated building in Bialystok was constructed (Figure 7). Over time, most of the new buildings in the city were built using the OW-T/67 system. It is estimated that by 1990 over 750 residential buildings containing 45,000 flats had been built in Bialystok, housing around 140,000 people—almost half the city's population [26].

(a) (b)

Figure 7. Prefabricated buildings in the "large-panel system" OW-T/67 style: (**a**) a 5-storey building, the most common in Poland, 3 Piastowska St., the first to be built in Bialystok in 1969; (**b**) an 11-storey building, 11 Mieszka I St., shown in 2021, source: authors—photo A. Turecki.

2.4. Methodology

The energy performance of a building is calculated in accordance with a national methodology [27], corresponding to the requirements of EP and R Directive 2010/31/EU and subsequent amendments [28,29]. This is an official tool used for designing and verifying the energy efficiency of new and modernized buildings in Poland and for calculating the actual energy performance of buildings.

The applied calculation method is based on PN-EN ISO 52000-1, 52003-1, 52010-1, 52016-1, and 52018-1 standards.

The energy performance is determined based on calculations of the typical energy consumption by the building's built-in systems, using the Audytor OZC 6.9 Pro software [30] with an implemented calculation model in accordance with the national methodology. According to the Directive 2010/31/EU, the performance is determined by the seasonal primary energy demands. In residential buildings, the calculated annual primary energy demands for heating, cooling, domestic hot water, and ventilation are calculated as follows (1):

$$Qp = \sum n \ Qp,n \ kWh/a \quad (1)$$

Here, Qp is the annual primary energy demand and Qp,n is the annual primary energy demand for the system.

The primary energy consumption for individual systems is calculated based on national primary energy input indices (Table 1) (2):

$$Qp,n = Qk,n \cdot wi,n + Eel,pom,n \cdot wel \ kWh/a \quad (2)$$

Table 1. Primary energy input factors.

Energy Source	W_i
Cogeneration coal	0.80
Cogeneration biomass	0.15
Cogeneration municipal waste	0.15
Fossil fuels	1.10
Grid electricity	3.0

Here, Qk,n is the annual final energy demand for the respective system, wi,n is the primary energy input factor for the system, Eel,pom,n is the annual electricity demand for the auxiliary equipment of the system, and wel is the primary energy input factor for the grid electricity according to KOBIZE for 2020 [31].

$$Eel,pom,n = \sum n \ qel,n \cdot t,el,n \cdot Af \cdot 10-3 \ kWh/a \quad (3)$$

Here, qel,n is the unit electric power demand to drive the auxiliary equipment in question, tel,n is the operating time of the appliance in question, and Af is the area of temperature-controlled space in the appliance's operating zone.

For systems supplied with district heating, the average primary energy input factor was calculated on the basis of data for the local district heating systems (3), along with calculation of the shares of individual fuels, the heat generation method, and national primary energy input factors [32] (Figure 8).

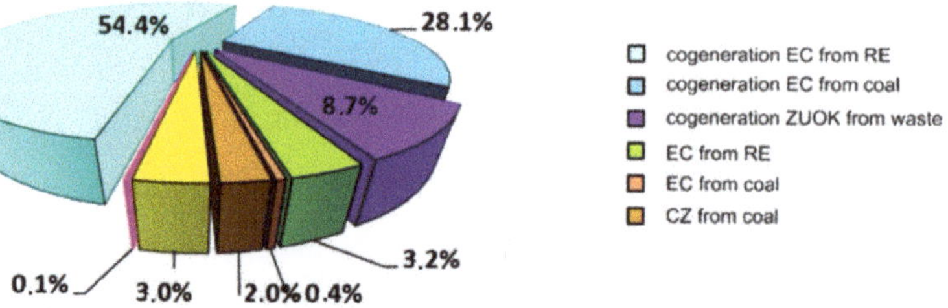

Figure 8. Heat structure in Bialystok district heating network depending on the source and generation technology—data published for design calculations in 2021, source: authors—prepared by M. Tur.

The specific CO_2 emissions were calculated as the sum of emissions from the energy consumption levels of the individual systems in the buildings related to the useful floor area of temperature-controlled buildings. CO_2 emissions were determined for individual fuel shares in the energy mix of the district heating network, taking into account emission factors based on measurement data collected by the distributor [33].

Emissions from electricity consumption were calculated based on the CO_2 emission factor of 719 kg CO_2/MWh of electricity from the national electricity grid for 2019 according to data from the National Center for Balance and Emissions Management [34]. The same CO_2 emission factors and shares of fuels in the heating network mix were assumed for all calculation variants using the latest data available for 2021.

Calculation Variants

Calculations were carried out for selected types of multifamily buildings that constitute significant proportions of the residential housing stock in the cities of Bialystok and Szczecin [35]. The buildings' energy consumption and CO_2 emissions were calculated for different building states.

For all variants, the same primary energy input and CO_2 emission indices were adopted, which are valid on the day of calculation.

Calculation variants:

Here, I is the original state, determined on the basis of archival documentation and regulations in force at the time of the building's construction, involving a thermal shield without a thermal insulation layer or highly insufficient natural ventilation, with air inflow through leaky windows and an assumed air tightness of n50 = 6.0 1/h (Table 2).

Table 2. Heat losses in the building at 3 Piastowska Street in Bialystok at the time of construction.

Heat Losses	kWh/a	%
Windows and doors	346542	37.2
Thermal shield (without windows)	270712	29
Heat for ventilation	314577	33.8
Total	931831	100

Here, II is the existing condition; the analyzed buildings, as well as the majority of the surveyed resources, were subjected to varying degrees of thermo-modernization over the years 1990–2020, covering in the vast majority of cases thermal shielding of the buildings, except for floors on the ground and internal walls; window replacements and modernization of the building's heating system; thermal insulation of the thermal envelope (R = 3.0–4.0 [$m^2 \cdot K/W$]); and airtight fitting of the windows, with an average coefficient of U = 1.5 [W/m^2 K]. The airtightness of the building was assumed to be n50 = 3.0 1/h from the average of measurement results from airtightness tests carried out on a sample of 22 residential buildings according to PN-EN 13829. The actual energy consumption readings from the meters in the investigated buildings were around 20% lower than the calculation results, due to insufficient flow of supply air after replacement of the old windows with airtight ones (Table 3).

Table 3. Heat losses in the building at 3 Piastowska Street in Bialystok—present state.

Heat Losses	kWh/a	%
Windows and doors	187892	32.3
Thermal shield (without windows)	142675	24.4
Heat for ventilation	252700	43.3
Total	583267	100

Supplementary retrofit options have been adopted:

Here, III represents the adaptation of all thermal shielding parameters with installation meeting the requirements of the EU national regulations. The adopted thermal insulation system has a thermal shield R = 4.0–6.0 [m^2·K/W], while the windows have a coefficient U = 0.9 [W/m^2 K] with a high level of tightness, along with natural ventilation with the provision of regulated air supply through window ventilators (Table 4).

Table 4. Heat losses in the building at 3 Piastowska Street in Bialystok—a modernization variant including adjustment of the building envelope to the present thermal insulation requirements.

Heat Losses	kWh/a	%
Windows and doors	138228	30.2
Thermal shield (without windows)	84831	18.6
Heat for ventilation	233979	51.2
Total	457038	100

Here, IV represents a variant involving the installation of a mechanical ventilation system with heat recovery allowing for reduced energy consumption when heating ventilation air [36] and improved hygienic conditions resulting from the air exchange in rooms. Windows in most buildings are fitted with PVC frames, without the installation of ventilation ventilators, while the flow of gravitational ventilation is inhibited by sealing the building envelope, leading to a reduction in the intensity of gravitational ventilation below hygienic requirements [37].

Mechanical ventilation systems are installed in installation shafts mounted on the external walls of staircases, with individual air handling units installed in staircases. The assumed total efficiency of the recuperation system is 49% and the building air tightness level recommended by the regulations is n50 = 1.5 1/h (Table 5).

Table 5. Heat losses in the building at 3 Piastowska Street in Bialystok—modernization variant including modernization of the building envelope and installation of a mechanical ventilation system with heat recovery.

Heat Losses	kWh/a	%
Windows and doors	138283	36.2
Thermal shield (without windows)	85051	22.2
Heat for ventilation	159170	41.6
Total	382504	100

Here, V represents the replacement of the heat source used for heating the building and the DHW heating from the existing connection to the district heating network with a system powered by highly efficient heat pumps with an efficiency COP = 4.0, with vertical ground probes as the source of energy. The district heating network in Bialystok is supplied with combustion heat, the modernization variant provides for the use of geothermal energy and includes the installation of a passive cooling system involving vertical ground probes.

Here, VI represents a variant involving the installation of a photovoltaic system on the building roofs. This system can be installed on flat roofs of multifamily buildings without major interference with the interior. An analysis of the possibility of installing PV systems on exemplary buildings shows that about 50% of the roof area can be used (Figure 9). For the PV calculation gains, a new method of assembling panels was used. The PV panels form pairs touching each other at the top, with a slight inclination of 15°, parallel to the edges of the roofs and with the azimuth pointing in the W–E direction. Although the yield of the individual panels is lower than with a south-facing orientation, this is compensated for by the smaller shading areas and the larger number of panels installed. The above mean that PV system produces slightly less energy but is more resistant to damage by strong winds. The production volume of PV systems (also with different settings) is calculated using the EU PV calculator [38]:

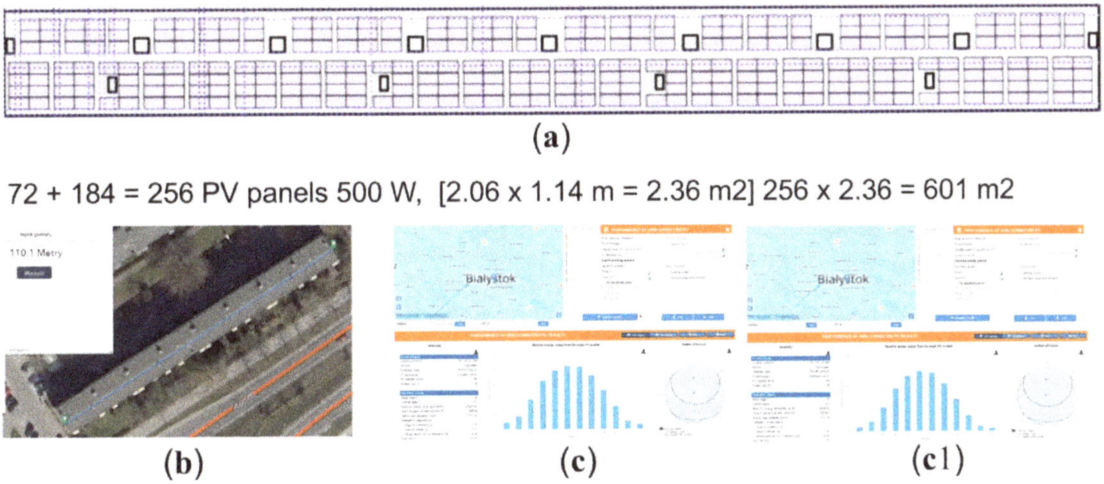

Figure 9. PV system on the roof of Piastowska 3: (**a**) plan of PV panels; (**b**) photo—Google Earth 2021; (**c**,**c1**) calculation of PV system energy production, source: authors—prepared by A. Turecki.

3. Results

The calculation charts (Table 6) show us the amount of energy consumed by the exemplary buildings. AUDYTOR software allows us to make a backward calculation, showing how much energy was consumed by each building at the time of completion for variant I (Table 7, Figure 10).

In previous years, nobody was particularly interested in this issue—energy was cheap, sources seemed to be unlimited, not many people were aware of the existence of CO_2, and nobody thought that it could affect global warming. However, in these data, in addition to information showing the technical advancement of the building, aspects related to building size dominate, which need to be separated out.

The energy demand coefficient per unit of usable floor area in one year (EK) omits the building size problems, meaning the data becomes clearer (Table 8, Figure 11). A retrospective analysis showed that in pre-WWII buildings in Szczecin and in those built just after WWII in Bialystok, the EK coefficient exceeded 400 kWh/m^2·a for variant I buildings. After the year 2000, the increased requirements for energy efficiency in the building regulations obliged building owners to undertake thermal improvements, which reduced the EK coefficient values to close to 108–141 kWh/m^2·a for variant II buildings. Unfortunately, works undertaken in the first decades of the 20th century do not meet the current requirements; they need to be continued, sometimes by replacing almost twenty-year-old materials with new ones. This will make it possible to reduce the EK values to 90–101 kWh/m^2·a.

Table 6. Summary calculations of different variants of energy retrofit for the Piastowska St. 3 building.

Building, Adress	**OWT/67—3 Piastowska St., Bialystok**					
Year of Build	**Heated Area (Useful area) of a Building [m²]**			**A/V**		
1968	**4794**			**0**		
Variant Number	**I**	**II**	**III**	**IV**	**V**	**VI**
	Primary Variant—As First Built	**Existing State—Modernisation of Windows, Walls, Roofs and Heating System (2007–2020 Standard)**	**Full Modernisatiuon of Building Envelope (2021 Standard)**	**+ Modernisation of Ventulation System with Heat Recovery**	**'+ Ground Heat Pump and Passive Cooling**	**'+ PV Installation (Roof Integrated)**
EU [kWh/m² a]	125	56	38	0	0	0
QH [kWh/a]	597528	[illegible]94	184266	0	0	0
Qk,H [kWh/a]	1047074	347827	238073	128407	35413	35413
Qk,V [kWh/a]	0	0	0	54567	27086	27086
Qk,W [kWh/a]	361059	225840	193777	193777	50754	50754
Qk,C [kWh/a]	0	0	0	0	18674	18674
QPV [kWh/a]	0	0	0	0	0	−72870
Qk, [kWh/a]	1408133	573667	431850	376751	131927	59057
EK [kWh/m² a]	**294**	**120**	**90**	**79**	**28**	**12**
QP, [kWh/a]	591416	240940	181377	158235	395781	177171
Change of energy demand Qk [%]	245%	100%	75%	66%	23%	10%
Total emission CO_2[kg/a]	**460499**	**187767**	**141736**	**131268**	**94857**	**42462**
Change of CO_2 emission [%]	245%	100%	75%	70%	51%	23%
CO_2 emission factor [kgCO_2/m² a]	**96**	**39**	**30**	**27**	**20**	**9**
residents number		**171**				
Emission CO_2 factor per resident [kgCO_2/pers a]	**2693**	**1098**	**829**	**768**	**555**	**248**

Table 7. Final energy demands (Qk (kWh/a)) for 6 analyzed technical modernization variants of the considered buildings.

Building	Final Energy Demand Qk [kWh/a]					
	I	II	III	IV	V	VI
2 Krasinskiego	894518	312338	224799	197737	65033	18973
3 Piastowska	1408133	573667	431850	376751	131927	59057
11 Mieszka I	3159420	1447473	983992	864972	324846	241896
27 Mickiewicza	1659884	813614	616918	538453	193155	130575
20 Janickiego	615562	175793	121342	114195	39578	−6292
36 Rostworowskiego—1 section	211925	65892	45398	45297	16340	2829
17 Brzozowskiego—1 section	166881	39732	33118	35691	13021	983

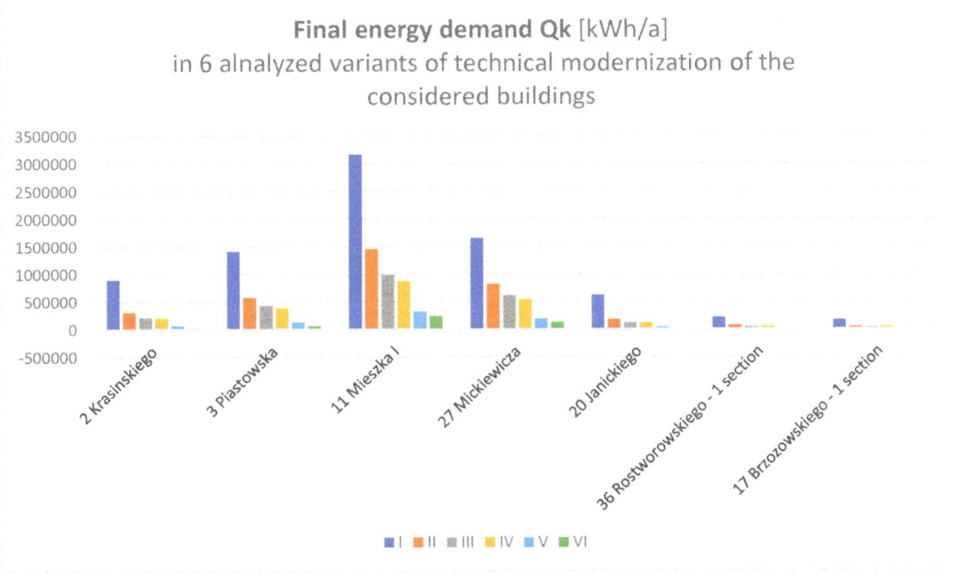

Figure 10. Final energy demands Qk (kWh/a) for 6 analyzed technical modernization variants of the considered buildings, source: authors—prepared by A. Turecki.

Table 8. Energy demands (Ek (kWh/m^2·a)) of 6 analyzed technical modernization variants of the considered buildings.

Building	Energy Demand EK [kWh/m^2·a]					
	I	II	III	IV	V	VI
2 Krasinskiego	403	141	101	89	29	9
3 Piastowska	294	120	90	79	28	12
11 Mieszka I	271	124	84	74	28	21
27 Mickiewicza	236	116	88	76	27	19
20 Janickiego	417	119	82	77	27	−4
36 Rostworowskiego—1 section	418	130	89	81	29	6
17 Brzozowskiego—1 section	455	108	90	85	35	3

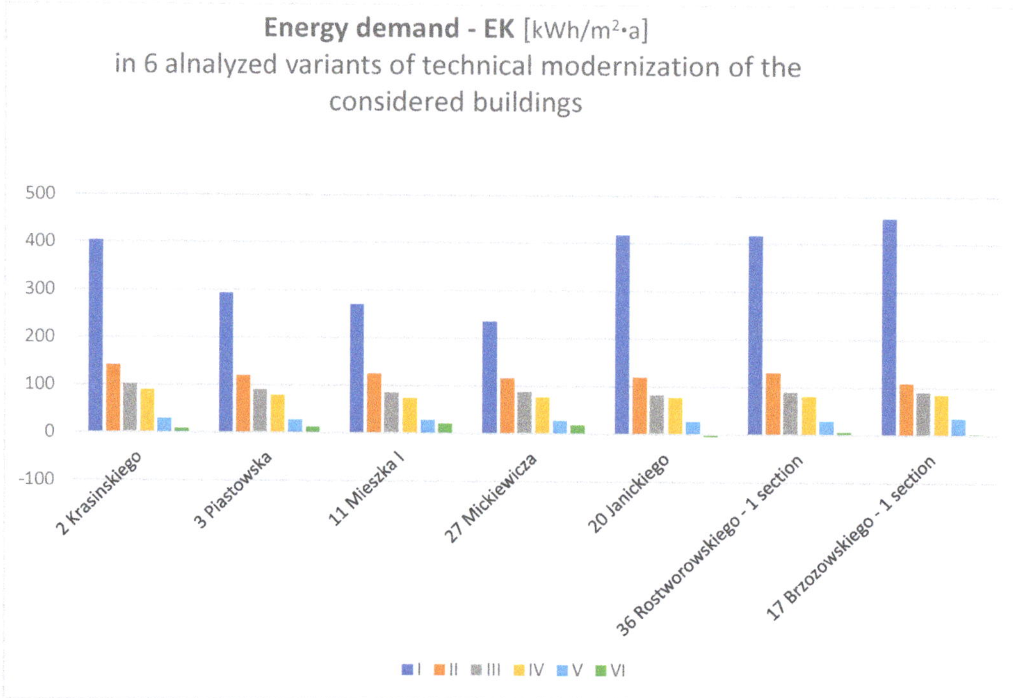

Figure 11. Energy demands (Ek (kWh/m^2·a)) of 6 analyzed technical modernization variants of the considered buildings, source: authors—prepared by A. Turecki.

The values shown in Figure 11 are still unsatisfactory. The current EU directive requires new buildings to be nearly zero-energy.

Therefore, it is necessary to introduce advanced technologies. The basic requirement, because we breathe and must have fresh air with appropriate humidity, is ventilation with recuperation (at the same time this is the most difficult one, because it requires the installation of additional systems inside the flats, meaning it may be easier to replace all installed systems in buildings that are already at thirty years old). Regarding recuperation, variant IV will allow the EK values to be reduced to 74–89 kWh/m^2·a (in Szczecin, recuperation systems have not been installed in buildings due to their historic status).

The next steps require the use of renewable energy sources. The use of a ground source heat pump (GSHP) gives significant results (variant V). The COP indicator showing the energy yield compared to the amount of energy needed to run the device in modern PCs is higher than 4. An additional advantage of using GSHP is the ability to cool buildings in summer by exploiting the "coolness" of the ground, involving minimal energy consumption (only the circulation pumps are run, while the HP compressor is switched off). A favorable aspect of the buildings is the large courtyard areas, which allow for the construction of many ground probes for the lower heat sources. The use of GSHP will significantly reduce the EK values to 27–35 kWh/m^2·a.

The last technology proposed is the photovoltaics (PV) variant (variant VI). This system generates electricity in the most versatile and necessary manner to power almost everything in modern buildings, especially when operating heat pumps. Paradoxically, PV systems in Poland are currently developing faster than all of the previously mentioned technologies. This technology is also the simplest to implement; it can be installed on the roofs of multifamily buildings without major interference with the interior. An analysis of the possibility of installing PV systems on the exemplary buildings showed that about 50%

of the roof area can be used (the prior application of ventilation systems with recuperation may increase this area). The variant considered assumes that the PV system is integrated with the grid, i.e., the grid plays the role of a battery, which returns energy at night and in winter months, when PV systems in Poland produce less energy. The use of PV areas will allow the EK values to be reduced to 4–21 kWh/m^2·a.

The consumption of energy from individual sources or combined heat and power (CHP) plants has so far usually been associated with CO_2 emissions. Reducing such gas emissions is one of the main objectives of the EU's "Fit For 55" program. Taking into account the combustion of different types of fuels at the combined heat and power plant in Bialystok and the KOBIZE data for Szczecin, the amounts of CO_2 emitted over 1 year were calculated for different renovation options for the analyzed buildings (Table 9, Figure 12).

Table 9. Total CO_2 emissions (kg/a) for 6 analyzed technical modernization variants of the considered buildings.

Building	Total Emission CO_2 [kg/a]					
	I	II	III	IV	V	VI
2 Krasinskiego	391958	102134	73732	68354	46758	13642
3 Piastowska	460499	187767	141736	131268	94857	42462
11 Mieszka I	1034056	477767	323124	301117	244203	173923
27 Mickiewicza	527388	242408	205726	190255	138879	93883
20 Janickiego	199503	40008	28222	33428	25883	0
36 Rostworowskiego—1 section	70602	14754	10376	13252	10475	2034
17 Brzozowskiego—1 section	54288	11130	7214	10423	8339	707

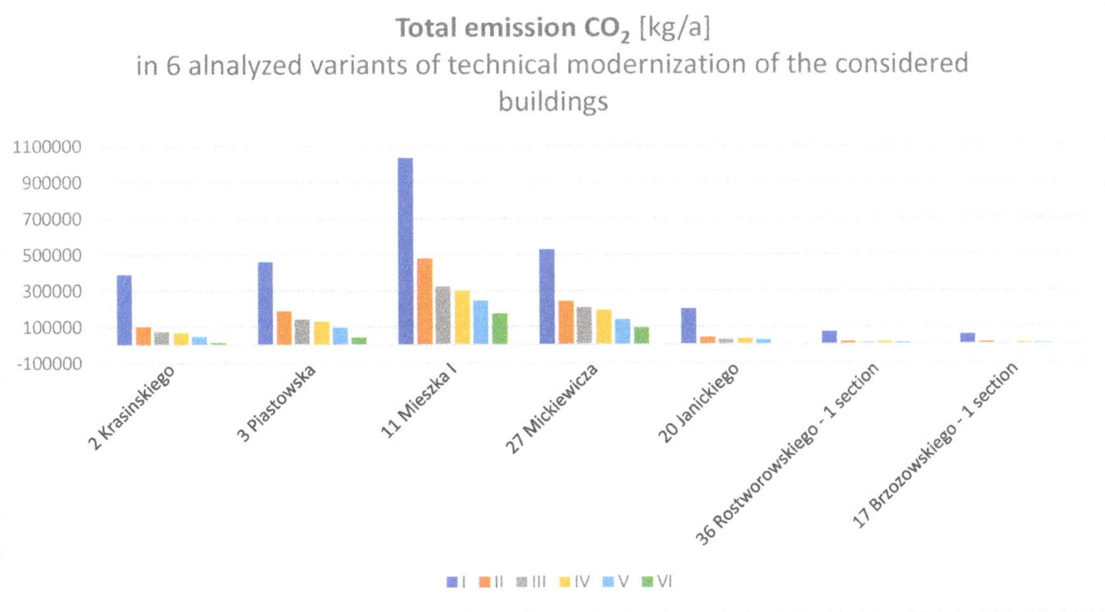

Figure 12. Total CO_2 emission (kg/a) for 6 analyzed technical modernization variants of the considered buildings, source: authors—prepared by A. Turecki.

The data show increases in CO_2 emissions for variant IV buildings in Szczecin. These increases result from the connection of the buildings to the municipal heating network when previous individual gas furnaces in flats have been removed, as required by law. Gas is an energy carrier that emits less CO_2 than heat supplied by the CHP plant in Szczecin, which burns mainly coal. Introducing a PV system to the 20 Janickiego St. building in Szczecin will result in zero CO_2 emissions.

As is the case for energy consumption, data on the total consumption should be supplemented with data showing the CO_2 emission coefficient per 1 m^2 in one year—this omits the size of the building and shows the levels of building emissions for particular retrofit variants (Table 10, Figure 13).

Table 10. CO_2 emissions ($kgCO_2/m^2 \cdot a$) for 6 analyzed technical modernization variants of the considered buildings.

Building	CO_2 Emission Factor [$kgCO_2/m^2 \cdot a$]					
	I	II	III	IV	V	VI
2 Krasinskiego	177	46	33	31	21	6
3 Piastowska	96	39	30	27	20	9
11 Mieszka I	89	41	28	26	21	15
27 Mickiewicza	75	34	29	27	20	13
20 Janickiego	135	27	19	23	18	0
36 Rostworowskiego—1 section	139	29	20	26	21	4
17 Brzozowskiego—1 section	148	30	20	28	23	2

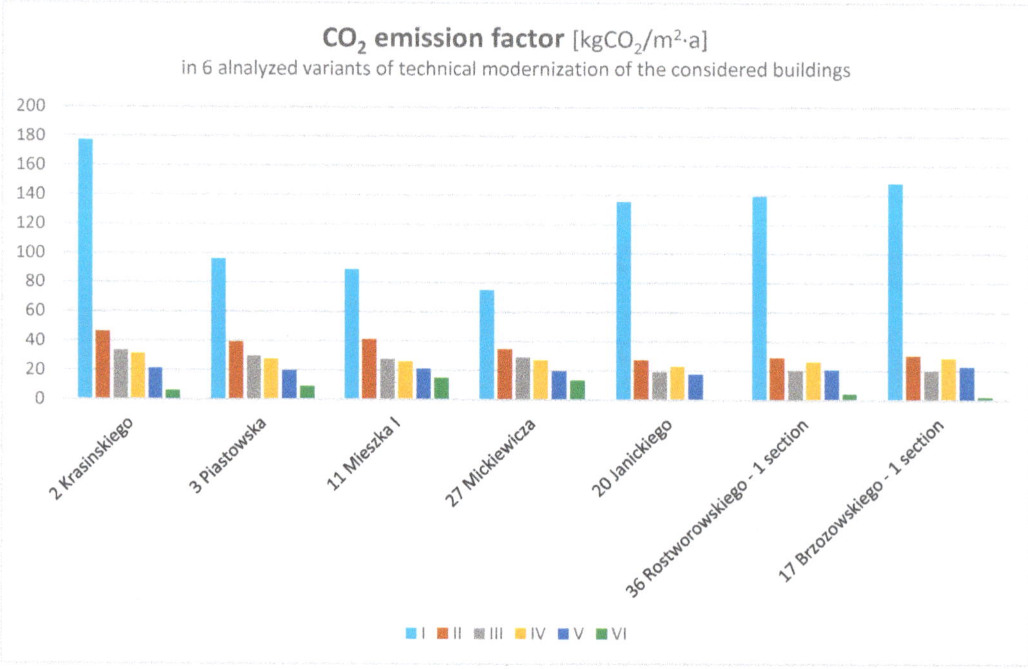

Figure 13. CO_2 emissions ($kgCO_2/m^2 \cdot a$) for 6 analyzed technical modernization variants of the considered buildings, *source: authors—prepared by A. Turecki.*

The very low emission values for variant VI buildings in Szczecin with the application of PV systems are due to the low height of the buildings—they are three storeys high, which means that a larger amount of PV panels can be used per 1 m^2 of flat space.

Each modernization variant has financial implications. In our case, this is the least certain element of the analysis. The current inflation rates, interest rates on borrowed loans, price turmoil in the market, volatility of energy prices, and in the case of some technologies, rapid increases in efficiency with declining unit prices and volatility of the subsidy policy make the financial analyses very uncertain. The AUDYTOR program also performs such analyses. The solutions with the shortest payback times are preferred. The calculations assume prices from 2021. The current prices are changing significantly. The current results for 2021 should be taken as illustrative (Table 11, Figure 14).

Table 11. Net costs for technical modernization variants of the considered buildings (euros). Note: T—total cost; III—envelope; IV—ventilation with recuperation; V—heat pump with GS; VI—PV.

Building	Nett Costs of Technical Modernization [Euro]				
	Total	III	IV	V	VI
2 Krasinskiego	390675	174303	39231	81949	95192
3 Piastowska	738344	330483	85350	172083	150427
11 Mieszka I	1377190	631761	190085	383763	171581
27 Mickiewicza	902017	395447	117829	209521	179220
20 Janickiego	330983	126951	41186	68829	94017
36 Rostworowskiego—1 section	132538	64641	16597	23682	27618
17 Brzozowskiego—1 section	102199	42940	14934	19645	24679

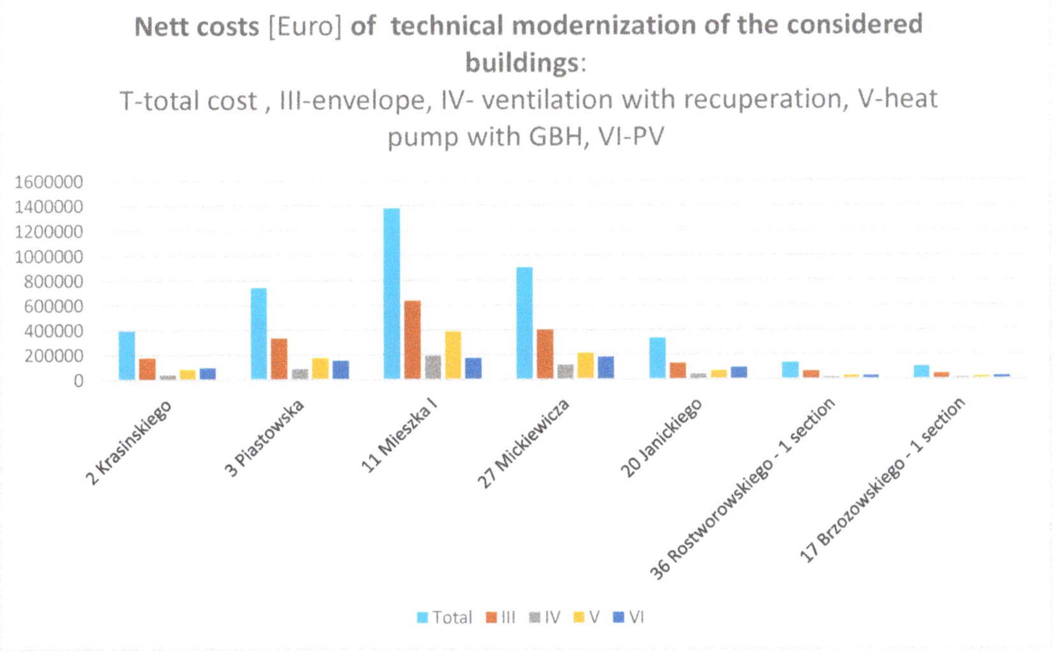

Figure 14. Net costs for technical modernization variants of the considered buildings (euros). Note: T—total cost; III—envelope; IV—ventilation with recuperation; V—heat pump with GS; VI—PV, source: authors—prepared by A. Turecki.

As for energy and CO_2 emissions, the results are presented as total values per 1 m^2 of a dwelling (Table 12, Figure 15). It is important to show these costs per capita, but such results may also be uncertain, as these numbers are constantly changing (Table 13, Figure 16).

Table 12. Net costs for technical modernization variants per 1 m^2 of usable area (euros/m^2) for the considered buildings. Note: T—total cost; III—envelope; IV—ventilation with recuperation; V—heat pump with GS; VI—PV.

Building	Nett Costs of Technical Modernization of 1 m^2 [Euro/1 m^2]				
	Total	III	IV	V	VI
2 Krasinskiego	176	79	18	37	43
3 Piastowska	154	69	18	36	31
11 Mieszka I	118	54	16	33	15
27 Mickiewicza	128	56	17	30	25
20 Janickiego	224	86	28	47	64
36 Rostworowskiego—1 section	261	127	33	47	54
17 Brzozowskiego—1 section	243	102	35	47	59

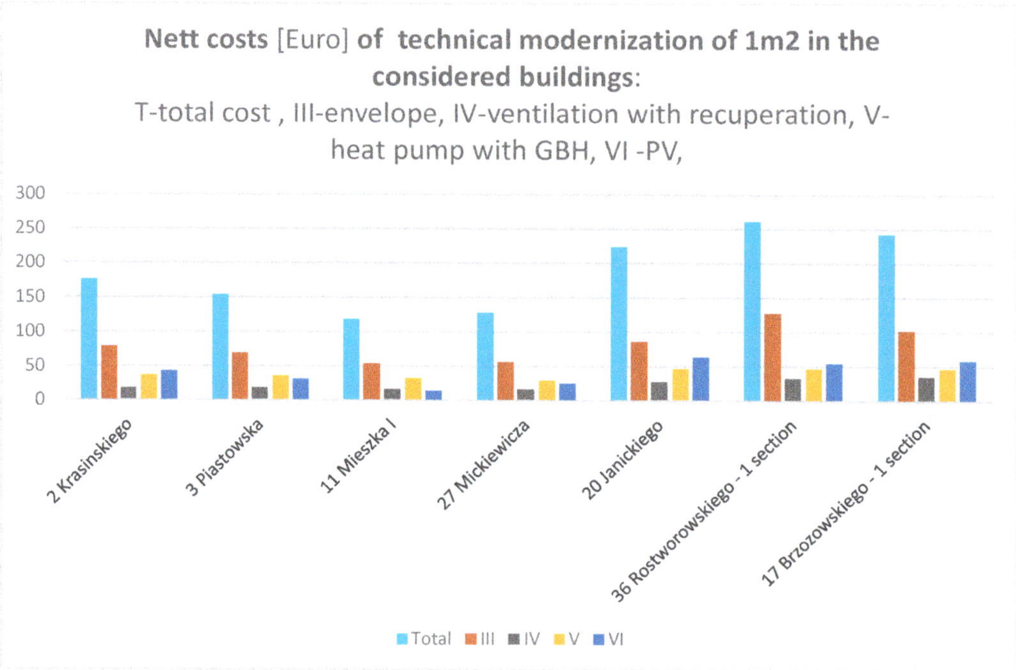

Figure 15. Net costs for technical modernization variants per 1 m^2 of usable area (euros/m^2) in the considered buildings. Note: T—total cost; III—envelope; IV—ventilation with recuperation; V—heat pump with GS; VI—PV, *source: authors—prepared by A. Turecki.*

Table 13. Net costs for technical modernization variants per inhabitant of the considered buildings (euros/person). Note: T—total cost; III—envelope; IV—ventilation with recuperation; V—heat pump with GS; VI—PV.

Building	Nett Costs of Modernization per Inhabitant [Euro/Person]				
	Total	III	IV	V	VI
2 Krasinskiego	6104	2723	613	1280	1487
3 Piastowska	4318	1933	499	1006	880
11 Mieszka I	3469	1591	479	967	432
27 Mickiewicza	4235	1857	553	984	841
20 Janickiego	4662	1788	580	969	1324
36 Rostworowskiego—1 section	5763	2810	722	1030	1201
17 Brzozowskiego—1 section	8517	3578	1244	1637	2057

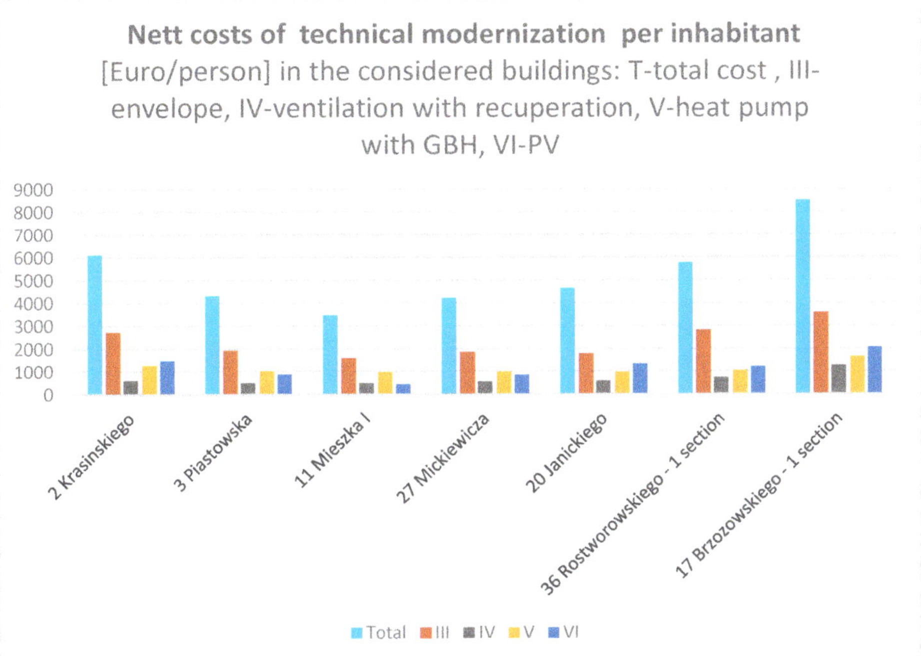

Figure 16. Net costs for technical modernization variants per inhabitant in the considered buildings (euros/person). Note: T—total cost; III—envelope; IV—ventilation with recuperation; V—heat pump with GS; VI—PV, source: authors—prepared by A. Turecki.

4. Discussion

The renovation of existing multifamily housing developments has been described in the scientific literature. As the articles come from different countries with different building traditions and climate zones, the conclusions of the published research sometimes differ significantly. It is advisable that recommendations be developed for all regions of Europe.

Adison et al. [39] presented the idea to extend the cost–benefit analysis approach in context of the financial returns from energy efficiency measures. They proposed an examination of factors such as the transaction costs and energy efficiency services. The approach we proposed states that the introduction of solutions based on renewable energy sources is necessary to achieve a 55% reduction in CO_2 emissions by 2030, even though the costs are significant.

Lee et al. [40], based on very a large research sample (16,158) of old houses, selected 11 key input variables that affect the heating energy consumption. They referred to the two main approaches involved in predicting building energy consumption: physics-based models and data-driven models.

Our approach is closer to the second method. We used energy consumption data provided by the owners of the analyzed buildings, information about the fuels burnt in the local CHP plant in Bialystok, population registers from the Municipal Office, and data collected in the building inventory. The energy performance was determined based on calculations of typical energy consumption by the building's built-in systems using Audytor OZC 6.9 Pro software with an implemented calculation model in accordance with the national methodology, corresponding to the requirements of EP and R Directive 2010/31/EU.

Lee's key variables were the roof U-value, roof area, wall U-value, floor area, floor U-value, year of completion, wall area, heating space area, boiler efficiency, window area, and window U-value.

In our research, we considered the same factors; in addition, the Audytor software requires the precise types and properties of the materials used in the building to be considered, including the location (by matching it with real climatic data and the geographical azimuths of the building to calculate the solar radiation gains), the number of users, and the building equipment, in order to calculate the internal energy gains. The electricity production from the PV systems was calculated using the EU PV calculator.

As a main factor, we considered the building envelope thermal insulation values. Jezierski et al. [41] examined the impacts of changes in these values on the energy demands, heating costs, and emissions. They developed mathematical models based on these dependences. The energy demands were assessed at three levels, corresponding to the maximum required values as approved by Polish law in 2014, 2017, and 2021. Jezierski defined the impact of changes in Polish building insulation standards in those periods, resulting in a ca. 27% decrease in heating energy demands. In of our estimations, the buildings were constructed in the years prior to energy efficiency standards, so increasing the insulation of the building envelopes in the oldest housing estates could even result in 100% increases in efficiency.

Jezierski et al. also took into account the different Polish building laws across different areas of Poland based on climate zones; the locations of buildings according to climate zones caused differences of up to 32.6% in energy demands. The Audytor OZC 6.9 Pro software calculations were more precise thanks to the inclusion of building azimuths in the calculation of solar gains.

Lu and Memari [42] presented a comparison of measurement methods for the thermal transmittance of building envelopes. These methods were highly precise.

According to estimations for objective 1 (differences between the actual and expected energy standards based on selected examples), it should be noted that the thermal insulation works performed so far need to be improved (variant III). Modern insulation standards are more stringent and another layer of insulation should be added to the walls and the roof. It would be advisable to replace the windows—the previous double-glazed ones had insulation values of U-value 1.3–1.6 W/m^2 K—with newer triple-glazed ones with a U-value of 0.8 W/m^2 K. This will reduce the energy consumption per 1 m^2 by approximately to 68–83% (the percentage reduction was calculated in relation to the existing state (variant II). The next step is to improve the building's ventilation system and replace the existing gravity ventilation with a mechanical system with an energy recovery function (variant IV). Despite the onerous nature of the work, which requires intervention inside the flats, there is a need to replace all installations, as they are already at the end of their useful life, and to introduce outdoor air filtration systems, as in many Polish cities smog levels often exceed acceptable standards and because gravitational ventilation systems works ineffectively on warm days with temperatures above 25 degrees Celsius. Increasingly frequent periods of hot summer weather will soon force the introduction of cooling systems in flats and ventilation by opening windows will become unacceptable. This stage of modernization will not reduce energy consumption very much—likely to 63–78% of current values—but it is necessary for hygiene and well-being reasons. Significantly, greater reductions in energy consumption will be achieved in variant V buildings through the use of heat pumps with GSHP ground probes for general heating and water heating. This technology will reduce energy consumption to 21–33% of current rates. In addition, it will enable low-cost passive cooling of flats in summer. The last technology variant (variant VI) involves the use of photovoltaics systems for electricity production. PV systems will be installed on the flat roofs of buildings. The scale of reduction will depend on the number of storeys—in lower 3–4-storey buildings, it will be higher at 0–6%, in 5-storey buildings it will be 10%, and in 11-storey buildings it will be about 17%. Of course, PV systems can also be mounted on the walls of buildings, but due to the complications involved, this possibility was not included in the study. However, the possibility of the future integration of photovoltaics with insulating elements—creating BIPV wall cladding panels—should be investigated, and work is also underway on thermo-photovoltaics systems that can generate electricity and heat, improving the system's efficiency via cooling.

Application of technology variants III–VI will reduce energy consumption levels in the analysed buildings by 83–100% compared to present levels (the building at 20 Janickiego St. would even become energy-positive).

According to objective 2 (relating to the CO_2 emissions connected with energy consumption, in particular modernization variants, under EU "Fit For 55" targets), the tools, procedures, and calculations used in the article are compatible with Polish building law standards. The novel method we introduced in the article is a reference for building energy efficiency level and CO_2 emission measurements, which is compatible with the goals of the EU's newest "Fit For 55" program. Additionally, we have proposed a procedure for estimating the modernization needs of communist era blocks of flats that can be applied to all such habitable resources throughout the country, along with estimations of future technical standards.

The Energy Forum Agency (VI 2021) states that heating buildings resulted in emissions in 1990 of 35.2 million tonnes of CO_2, while by 2018 this had been reduced to 35.1 million tonnes—a reduction of just 0.2% [43]. The "Fit For 55" target will, therefore, require a reduction of 19.1 million tonnes of CO_2 by 2030. The restructuring of heating systems in single-family houses, which has been occurring for several years—including the Polish "Clean Air" Program—seems to be failing, so it is difficult to expect success in the sector of single-family housing. It must be stated that the majority of measures to reduce greenhouse gas emissions must be carried out in multifamily buildings and in public sector buildings.

The assumptions of replacing coal with gas in the energy mix also need to be revised, as the current perturbations in the gas supply market caused by Russia's war with Ukraine show.

In order to achieve the overall reduction target of 55% in this sector, much higher individual levels of reduction must be achieved, especially as this 55% is only an intermediate target for 2030, with the final target being climate neutrality by 2050. This means in practice the implementation of zero-carbon architecture based on the use of renewables and clean sources of energy.

The calculations (in relation to the existing state—variant II) show that retrofitting a building's envelope (variant III) will reduce $CO_2/m^2{\cdot}a$ emissions to 65–85% of the present level, introducing mechanical ventilation with recuperation (variant IV) will reduce emissions to 63–78% of the present level (no recuperation was proposed in Szczecin due to the status of historical monuments), using GSHP (variant V) will reduce emissions to 46–57% of the present level (except for buildings in Szczecin with smaller reductions), and adding PV systems on roofs (variant VI) will reduce emissions to 0–39% of the present level.

To meet the "Fit For 55" target of reducing CO_2 emissions by 55% compared to the 1990 levels, both improvements in building envelope insulation and the use of available renewable energy technologies are required.

Examining objective 3 (including an analysis of the results with reference to the income level of the population to answer the question of whether the modernization of all such housing in Poland is possible with the residents' own resources, and if not to what extent it is necessary to provide external funds), we conclude that the costs of technical retrofitting per 1 m^2 with the use of all available technical means in the analysed buildings, depending on their historical status and size, are as follows: in lower 3-storey historical buildings in Szczecin, this will be highest at 224–261 E/m^2; in 4-storey traditional construction buildings such as 2 Krasinskiego St. this will be 176 E/m^2; in 5-storey prefabricated buildings such as 3 Piastowska St. this will be 154 E/m^2; and in large 11-storey buildings this will be lowest at 118–128 E/m^2.

These figures need to be recalculated according to the number of inhabitants in order to show the level of potential financial involvement. The calculated structural costs with the use of all available technical means in the analyzed buildings per 1 inhabitant are similar to the costs per 1 m^2. In the analyzed buildings, they are as follows: 2 Krasinskiego St.—6104 E/person; 3 Piastowska St.—4318 E/person; 11 Mieszka I St.—3469 E/person; 27 Mickiewicza St.—4235 E/person.

Unfortunately, no extracted data were extracted on the families and the occupational activities of the residents, some of whom may be pensioners. We also do not know how big the flats are that they occupy.

We can only assume with a large margin of error that the average net salary in Poland is currently around 900 E, so the costs of the abovementioned retrofitting per person would be at the level of 4–7 monthly net salaries. However, it should be assumed that not all people work and some receive low pensions, and probably a significant part of the population would not be able to cover the costs of retrofitting.

The renovation project should be subsidized by the government and grants from the EU "Renovation Wave" program.

5. Conclusions

1. In multifamily buildings, we should achieve higher levels of CO_2 reductions than the expected 55% target of "Fit For 55" program because the rate of reduction for single-family housing will be probably much lower, especially as this 55% is only an intermediate target for 2030, and the final target is climate neutrality by 2050.
2. CO_2 reductions of over 55% are possible by retrofitting building envelopes and introducing three available and proven technologies: recuperation (in buildings with historical monument status, this will not be advisable), renewable energy sources (e.g., heat pumps extracting heat from the ground), and photovoltaics.
3. The predicted costs of this retrofitting project per 1 m^2 are as follows: in lower 3-storey historical buildings the costs are estimated at 300 E/m^2; in 4-storey traditional construction buildings at 200 E/m^2; in 5-storey prefabricated buildings at 160 E/m^2; and in large 11-storey buildings at 120–150 E/m^2.
4. The predicted costs of this retrofitting project per inhabitant, depending on the building, are ca. 4000–6000 E.
5. A significant percentage of the inhabitants will likely not be able to cover such costs, and renovations should be subsidized by the government and grants from the EU "Renovation Wave" program.

The crucial conclusion of the article is that modernist housing developments, such as those built in Szczecin in 1918–1925 and in Bialystok in 1950–1990, represent a significant portion of existing multifamily buildings. Some of these buildings have architectural heritage, although they are substandard in many aspects; functionally, aesthetically, technically, and ecologically they do not meet current energy standards. Research into improving their envelopes, into modern ventilation and heating technologies, and into photovoltaics systems should be pursued. Renovations can include the use of renewable energy sources. The well-being of residents is also important, meaning the quality of apartments should be increased. The aim of the research was to analyze the renovation options in terms of energy efficiency, CO_2 emissions, and basic costs of the works.

The simplified energy calculation method was used to check the present buildings' energy demands to compare them with retrofitting results. We considered several retrofitting possibilities—minimal retrofitting fulfilling current standards, medium retrofitting with changes to the heating sources, and maximal retrofitting with the use of photovoltaics systems. The results show that without EU financial aid, which will soon be introduced under the "Renovation Wave" program, such modernization will be difficult, and CO_2 emissions reductions of 55% of the 1990 levels by 2030 will be impossible.

Author Contributions: Conceptualization, A.T. and M.T.; methodology, A.T. and M.T.; software, M.T.; formal analysis, K.J.; investigation, A.T., M.T. and P.F.; resources, M.T., P.F. and K.J.; writing—original draft preparation, A.T., M.T., B.C. and P.F.; writing—review and editing, A.T. and B.C.; visualization, M.T. and A.T.; supervision, K.J. All authors have read and agreed to the published version of the manuscript.

Funding: This research was co-funded under projects no. WZ/WA-IA/1/2020 and WZ/WA-IA/2/2020 at the Bialystok University of Technology, financed by a research subsidy from the Polish Ministry of Education and Science, and at the West Pomeranian University of Technology in Szczecin, financed by a research subsidy from the Polish Ministry of Education and Science.

Institutional Review Board Statement: Not applicable.

Informed Consent Statement: Not applicable.

Data Availability Statement: The following are available on demand at adamturecki@poczta.onet.pl: Multifamily blocks of flats in Szczecin and Bialystok.xls: energy demands and CO_2 emissions; PV calculations; costs of modernization; calculations and charts.

Acknowledgments: The authors would like to thank to City Hall Bialystok, the Adam Mickiewicz Housing Cooperative, NIBE-BIAWAR Bialystok Ltd., CORAL, W. Perkowski, J. Perkowski, Choroszcz, BARTOSZ Co. Bialystok, and the Municipal Archive in Szczecin for providing data and calculations for this article.

Conflicts of Interest: The authors declare no conflict of interest. The funders had no role in the design of the study; in the collection, analyses, or interpretation of data; in the writing of the manuscript, or in the decision to publish the results.

References

1. Zangheri, P.; Castellazzi, L.; D'Agostino, D.; Economidou, M.; Ruggieri, G.; Tsemekidi-Tzeiranaki, S.; Maduta, C.; Bertoldi, P. *Progress of the Member States in Implementing the Energy Performance of Building Directive*; Publications Office of the European Union: Luxembourg, 2021; p. 7.
2. Lechner, N. *Heating, Cooling, Lighting: Sustainable Design Methods for Architects*, 4th ed.; John Wiley & Sons: Hoboken, NJ, USA, 2014; p. 13.
3. IPCC. Summary for Policymakers. In *Climate Change 2021: The Physical Science Basis. Contribution of Working Group I to the Sixth Assessment Report of the Intergovernmental Panel on Climate Change*; Masson-Delmotte, V., Zhai, P., Pirani, A., Connors, S.L., Péan, C., Berger, S., Caud, N., Chen, Y., Goldfarb, L., Gomis, M.I., et al., Eds.; Cambridge University Press: Cambridge, UK, 2021; pp. 15–18.
4. Lee, K. *Encyclopedia of Energy-Efficient Building Design: 391 Practical Case Studies*; Environmental Design and Research Center: Berkeley, CA, USA, 1977; pp. 573–574. Available online: https://books.google.pl/books?id=H4mEKu4jD6sC&pg=PA573&lpg=PA573&dq=Solar+1+MIT&source=bl&ots=KjDJwH8jr&sig=ACfU3U1mLobzktCe7ldkoA2RLYItqE_SjQ&hl=pl&sa=X&ved=2a (accessed on 22 February 2022).
5. Korsgaard, V.; Esbensen, T. Dimensioning of the solar heating system in the zero energy house in Denmark. *Sol. Energy* **1977**, *19*, 195–199.
6. Stahl, W.; Goetzberger, A.; Voss, K. *Das Energieautarke Solarhaus, Mit der Sonne Wohnen*; Muller, C.F., Ed.; Verlag: Heidelberg, Germany, 1997; pp. 53–57.
7. Architecture 2030. Available online: https://architecture2030.org/the2030challenges/ (accessed on 25 February 2022).
8. RenovationWave EU. Available online: https://energy.ec.europa.eu/topics/energy-efficiency/energy-efficient-buildings/renovation-wave_en (accessed on 25 February 2022).
9. Seligman, M.E.; Flourish, P. *A Visionary New Understanding of Happiness and Well-Being*; John Murray Press: London, UK, 2011.

10. Pritzkerprize. Available online: https://www.pritzkerprize.com/laureates/anne-lacaton-and-jean-philippe-vassal (accessed on 28 March 2022).
11. Turecki, A. *Untapped Potential of Loggias and Balconies Glazing*; Technical Transactions 11; CUT: Bloemfontein, South Africa, 2011; pp. 205–2011.
12. Lucchi, E.; Delera, A.C. Enhancing the Historic Public Social Housing through a User-Centered Design-Driven Approach. *Buildings* **2020**, *10*, 159. [CrossRef]
13. Dawidowski, R.; Długopolski, R.; Szymski, A. *Architektura Modernistyczna lat 1928–1940 na Obszarze Pomorza Zachodniego*; Walkowska Wydawnictwo: Szczecin, Poland, 2001; pp. 53–65, 128–180, 222–229.
14. Kuhn, G. *Citybildung und Dezentralisierung—Urbane Wohnprojekte in der Weimar Republik in Stadtwohnen. Geschichte-Städtebau-Perspektiven*; Harlander, T., Ed.; Wüstenrot Stiftung: München, Germany, 2007; pp. 184–195.
15. Hitchcock, H.-R.; Johnson, P. *The International Style*; Museum of Modern Art: New York, NY, USA, 1932.
16. Davies, C. *A New History of Modern Architecture*; Laurence King Publishing: London, UK, 2017; pp. 96–135.
17. Ciełątkowska, R.; Lorens, P. *Architektura i Urbanistyka Osiedli Socjalnych Gdańska Okresu XX-Lecia Międzywojennego*; Wydział Architektury Politechniki Gdańskiej: Gdańsk, Poland, 2000; pp. 5–46, 106–185.
18. Lampugnani, V.M. *Die Stadt im 20. Jahrhundert. Visionen, Entwürfe, Gebautes, Band I*; Verlag Klaus Wagenbach: Berlin, Germany, 2011; pp. 279–361.
19. Cohen, J.-L. *The Future of Architecture. Since 1889*; Phaidon: London, UK, 2012; pp. 176–199.
20. Durth, W.; Sigel, P. *Baukultur. Spiegel Gesellschaftlichen Wandels*; Jovis Verlag: Berlin, Germany, 2009; pp. 124–277.
21. Sołtysik, M.J. *Social House Building in Gdynia during the Interwar Period: Issues of Protection and Revitalization in International Conference Social Housing in the 1st Half of XXth Century in Europe: Problems of Preservation & Revitalization*; Wydział Architektury Politechniki Gdańskiej: Gdańsk, Poland, 2000; pp. 265–278.
22. Kleihues, J.P. *750 Jahre Architektur und Städtebau in Berlin. Die Internationale Bauausstellung Berlin 1987 im Kontext der Baugeschichte Berlins*; Verlag Gerd Hatje: Berlin, Germany, 1987; pp. 153–182.
23. Voigt, W.; Deschermer, D.; Cachola, S.P. *New Human, New Housing: Architecture of the New Frankfurt 1925–1933*; Dom publishers: Berlin, Germany, 2019; pp. 104–135.
24. Basista, A. *Betonowe Dziedzictwo*; Wydawnictwo Naukowe PWN: Warszawa, Poland, 2001; p. 85.
25. Basista, A. Dlaczego Czeka nas Odnowa Osiedli Zwanych Blokowiskami. In Proceedings of the Współczesna architektura mieszkaniowa: II Konferencja Naukowa Mieszkanie XXI wieku, Augustów, Poland, 26–28 May 1999; pp. 3–5.
26. Kłopotowski, M. *OW-T—Concrete Potential*; Architecturae et Artibus: Augustów, Poland, 2009; p. 4.
27. Regulation of the Minister of Infrastructure and Development of 27 February 2015 on the Methodology of Determining the energy Performance of a Building or Part of a Building and Energy Performance Certificates. Dz.U. 2015 pos. 376. Available online: https://isap.sejm.gov.pl/isap.nsf/DocDetails.xsp?id=WDU20150000376 (accessed on 3 March 2022).
28. Directive 2010/31/Eub of the European Parliament and of the Council of 19 May 2010 on the Energy Performance of Buildings. Available online: https://eur-lex.europa.eu/legal-content/PL/TXT/?uri=CELEX:32010L0031 (accessed on 3 March 2022).
29. Directive 2018/844 of the European Parliament and of the Council of 30 May 2018. Available online: https://eur-lex.europa.eu/legal-content/pl/TXT/?uri=CELEX%3A32018L0844 (accessed on 3 March 2022).
30. SANKOM. Available online: http://pl.sankom.net/programy/audytor-ozc (accessed on 4 March 2022).
31. Emission Factors of CO_2, SO_2, NO_x, CO and Total Dust for Electricity Based on Information Contained in the National Database on Emissions of Greenhouse Gases and Other Substances for 2019. National Centre for Balancing and Emissions Management, 2021. Available online: https://www.kobize.pl/uploads/materialy/materialy_do_pobrania/wskazniki_emisyjnosci/Wskazniki_emisyjnosci_grudzien_2020.pdf (accessed on 4 March 2022).
32. System Characteristics as at 01.01.2021—District Heating Network of ENEA Ciepło Sp. z o.o. Available online: https://www.enea.pl/pl/grupaenea/o-grupie/spolki-grupy-enea/enea-cieplo/system-cieplowniczy (accessed on 7 March 2022).
33. Regulation of the Minister of Infrastructure and Development of 27 February 2015, pos. 376, Unit ventilation air flow for the residential part Vve,1,s = 1.152 $m^3/h * m^2$, for staircases Vve,1,s = 0.792 $m^3/h * m^2$. *J. Laws* **2015**.
34. Fuel Structure and Emissions in 2020. ZUOK Bialystok. Data: ENEA Ciepło sp z.o.o: Source 1: Heat Plant West: Year 2018—92.58 kg CO_2/GJ; Year 2019—83.51 kg CO_2/GJ; Year 2020—71.45 kg CO_2/GJ, Source2: Bialystok CHP Plant: Year 2018—53.22 kgCO_2/GJ; Year 2019—49.87 kg CO_2/GJ; Year 2020—36.16 kg CO_2/GJ. Available online: http://www.lech.net.pl/resource/file/download-file/id.2016 (accessed on 10 March 2022).
35. Review of Regulations Specifying Minimum Requirements for the Energy Performance of Buildings—Exertion. Prof. Dr. Eng. Tadeusz Tatara, Cracow University of Technology, Malopolska Energy Saving Building Centre. p. 79. Available online: https://www.gov.pl/documents/33377/436740/EkspertyzaMCBE2017.pdf (accessed on 10 March 2022).
36. *PN-EN 13829*; Thermal Performance of Buildings—Determination of air Permeability Air Permeability of Buildings—Pressure Measurement Method Using a Fan. PKN: Warszawa, Poland, 2002.
37. Regulation of the Minister of Infrastructure on Technical Conditions to Be Met by Buildings and Their Location of 12 April 2002. Available online: http://isap.sejm.gov.pl/isap.nsf/download.xsp/WDU20190001065/O/D20191065.pdf (accessed on 11 March 2022).
38. EU PV Calculator. Available online: https://re.jrc.ec.europa.eu/pvg_tools/en/tools.html (accessed on 20 March 2022).
39. Adisorn, T.; Tholen, L.; Thema, J.; Luetkehaus, H.; Braungardt, S.; Huenecke, K.; Schumacher, K. Towards a More Realistic Cost–Benefit Analysis—Attempting to Integrate Transaction Costs and Energy Efficiency Services. *Energies* **2021**, *14*, 152. [CrossRef]

40. Lee, S.; Cho, S.; Kim, S.-H.; Kim, J.; Chae, S.; Jeong, H.; Kim, T. Deep Neural Network Approach for Prediction of Heating Energy Consumption in Old Houses. *Energies* **2021**, *14*, 122. [CrossRef]
41. Jezierski, W.; Sadowska, B.; Pawłowski, K. Impact of Changes in the Required Thermal Insulation of Building Envelope on Energy Demand, Heating Costs, Emissions, and Temperature in Buildings. *Energies* **2021**, *14*, 56. [CrossRef]
42. Lu, X.; Memari, A.M. Comparison of the Experimental Measurement Methods for Building Envelope Thermal Transmittance. *Buildings* **2022**, *12*, 282. [CrossRef]
43. Forum Energii. Ile Ważą Emisje CO_2 w Budynkach i Transporcie? 2021. Available online: https://www.forum-energii.eu/public/upload/artcles/files/46.%20Ile%20ważą%20emisje%20CO2%20w%20budynkach%20i%20transporcie%20fina\T1\l.pdf (accessed on 14 March 2022).

MDPI
St. Alban-Anlage 66
4052 Basel
Switzerland
Tel. +41 61 683 77 34
Fax +41 61 302 89 18
www.mdpi.com

Energies Editorial Office
E-mail: energies@mdpi.com
www.mdpi.com/journal/energies

www.ingramcontent.com/pod-product-compliance
Lightning Source LLC
Chambersburg PA
CBHW062045220926
51850CB00010B/2670